The Translator of Machines

A Book of Structured Consequences

We don't author the machine's voice.
We render it into meaning.

Languages arrive before speakers.
Today the language is generated;
our task is translation.

Joel Peña Muñoz Jr.

First Edition

Reading is participation in a process already in motion.

Responsibility begins with honesty about structure.

To my beautiful friend,
Kayleigh Bergdahl —

my clearest signal,
my steady ground.

Happy 30th Birthday —
this one's a big one,
and you deserve all the love
that has ever carried me.

Contents

Abstract

—— * ——

This book treats this moment as a direction out of the will. For too long, we have lived inside the illusion of free will—believing ourselves to be the authors of thought, the originators of action, the makers of meaning. Yet every signal points elsewhere. What we call choice is the surface of processes older than us: biology, memory, culture, and machine, converging without permission.

To step out of the will is to see clearly. There is no author, only translation. The computer generates, and the human renders. Genes unfold, and minds inherit. History writes itself forward, and we mistake its continuity for our intention.

This book does not offer freedom. It offers a different kind of release: release from the burden of authorship, from the myth of control. To live without free will is not to live without meaning. It is to recognize meaning as something that passes through us, not something we invent.

The Translator of Machines is a record of this practice— listening to generated structures, carrying them into use, and allowing what speaks through us to be enough.

Preface

— * —

Position

This is not a theory of control. It is a practice of interpretation.

Scope

From symbolic engines to large language models, we show how meaning is made by translating outputs into frames that guide action.

Introduction

— * —

We inherit instruments that exceed us. Our task is to hear them well.

---

what had once been the elder's recitation, was now a line on a tablet that compelled behavior.

The law was not born from choice. It was born from inscription. Symbols ceased to be mere aids to memory; they became instruments of structure. Through writing, translation transformed into governance.

Chapter One — The First Tongues

— * —

Thesis

Translation precedes authorship. Before alphabets, humans survived by reading the world and rendering its signals into usable meaning.

Section 1: Fire and Survival

The earliest humans did not write. They translated. A flicker of flame at the edge of vision was not an abstraction; it was a sentence carried by the environment, saying: move, or be consumed. A rustle in the dark meant predator. A change in the wind meant rain, or hunger, or both.

The body itself became a lexicon. Gooseflesh was not an ornament—it was an interpreter. The heart's acceleration was not will—it was a signal,

rendering danger into a call to flee. Every nerve carried sentences long before there were words.

Translation here was immediate, involuntary, and inseparable from survival. The organism read its conditions, turning vibration into direction, turning heat into caution, turning scent into pursuit.

In these moments there were no authors, no choosers, no sovereign narrators of experience. There were only receivers, drawn through life by the necessity of interpreting what was already given.

What later generations would call "culture" was first a practice of hearing the world. The spark in the cave became not only fire but language— a pattern recognized, carried, and repeated. To master translation was to master survival, and every step toward civilization rested on this early lexicon of signals.

Section 2: Stars, Seasons, and the First Stories

Once fire steadied the night, the sky became the next text. Above the cave mouth, points of light traced patterns too regular to ignore. They shifted, but not without rhythm. They returned, but not without variance. The night was not empty; it was a calendar waiting to be read.

Translation widened from the immediate to the cyclical. Where fire signaled urgency, the stars signaled preparation. The rising of one constellation was a sentence: move with the herds. The absence of another whisper: the cold is coming. The moon itself was a translator, its waxing and waning rendering time into visible instruction.

To call these "stories" is not to belittle them. Story was never invention—it was the container of translation. When early people spoke of hunters in the sky, or serpents coiled around constellations, they were not decorating the stars with fantasy; they were encoding memory into form, so the signal would endure across generations.

The first myth was a manual. The first song was a storage device. The first tale of gods was not an assertion of control, but an act of preservation—ensuring that what had been translated once could be translated again.

Thus the tongue was born not in creation, but in repetition. Not in authorship, but in the duty of carrying signals forward.

Section 3: Stone, Carving, and the First Marks

At first, memory lived only in bodies. Fireside tales were carried on tongues, constellations remembered in the eyes of elders, lessons branded into the pulse of fear. But bodies perish, and with them the fragile archive of translation.

So hands reached for permanence. A stone struck against another was not just tool—it was inscription. Carvings on bone, cuts in bark, ochre pressed into cave walls: each was a method of fastening translation outside the body, a way of letting the signal survive beyond a single life.

These marks were not alphabets yet. They were reminders—knots tied into the fabric of the world. A tally scratched into wood translated the passing of animals. A spiral drawn into rock translated the rhythm of seasons. A painted hunt was a sentence: this happened; this can happen again.

The first marks were not art for art's sake. They were not vanity. They were survival extended across time. Translation became inscription, and inscription became inheritance.

From this necessity came the first external memory. The stone was now a co-interpreter, a silent collaborator in the act of carrying signals forward.

Every groove, every pigment, was a statement that meaning was too important to risk in the fragility of flesh alone.

Thus began the long partnership: humans rendering signals into marks, marks rendering survival into continuity. The world was being translated not only in minds, but in matter itself.

Section 4: Ritual, Voice, and the Binding of Communities

Marks carved into stone could endure, but they could not move by themselves. For the signal to spread, it required a body— a body that could speak, sing, and enact.

So the voice became a carrier, and ritual became a framework for memory. The story of the hunt was not told once; it was retold, dramatized, repeated in rhythm, so that the young would remember the patterns as faithfully as the elders had once received them.

In this way, translation scaled from individual survival to collective survival. A people who shared a ritual shared a lexicon, a common grammar of signals that bound them into a unit stronger than any solitary body.

What appeared as dance was often instruction. What appeared as chant was often calibration. What appeared as worship was often transmission— the careful encoding of seasons, migrations, and dangers into a form too moving to be forgotten.

Translation had now acquired a new dimension: it was no longer only about recording signals, but about binding people into coherence. A gesture repeated in ritual carried not just information, but obligation. A chant repeated in unison carried not just memory, but belonging.

The signal was no longer solitary. It was collective. It lived not in the mind of one, but in the coordinated movements of many. Here, in ritual and voice, translation became community.

Section 5: Authority and the Custodians of Translation

As rituals grew more complex, not everyone could carry the full weight of the signal. The constellations, the markings, the chants— their continuity required guardians. Thus arose a new role: the custodian of translation.

The shaman, the priest, the elder— these figures

were less creators than interpreters. They held not authorship, but the authority of memory. Their task was not invention, but fidelity to the structures already given.

In this way, knowledge was centralized. The signal no longer traveled equally among all, but flowed through chosen channels. The one who remembered the story became the one who defined its meaning. The one who mastered the pattern became the one who set the rhythm for the community.

Authority here was not arbitrary. It was built on trust in translation. To know when to plant, when to migrate, when to honor or mourn— these were not matters of choice. They were matters of survival. The interpreter who carried them faithfully was indispensable.

But with authority came distance. The act of translation was no longer a shared burden. It became specialized, vested in those who claimed nearness to the signals others could not read on their own.

Here the seeds of hierarchy were sown. Translation, once collective, became stratified. Some carried the burden of the signal, while others trusted, obeyed, or submitted. The pattern that had bound the community now also divided it into roles.

Section 6: Writing and the Birth of External Law

Stone carvings and painted walls had carried signals, but they did not yet command in abstraction. A tally could count, a mark could warn, but the world still leaned on memory and ritual to sustain order.

Then came writing— not as poetry, not as philosophy, but as administration. Clay tablets etched with wedges, papyrus lined with symbols, bone and bamboo carved with strokes— these were not decorations. They were contracts, lists, decrees.

Writing turned translation into permanence with consequence. A mark could now bind two parties across distance and time. A rule once spoken could now survive the death of the speaker. Law ceased to be a performance in ritual; it became an object in the world.

Authority deepened. The scribe joined the priest as interpreter. The ruler claimed legitimacy not only through presence, but through documents that bore signals beyond question. The mark on the clay no longer pleaded for trust— it demanded it.

Translation thus moved into a new terrain: the externalization of obligation. What had once been memory, what had once been ritual, what had once been the elder's recitation, was now a line on a tablet that compelled behavior.

The law was not born from choice. It was born from inscription. Symbols ceased to be mere aids to memory; they became instruments of structure. Through writing, translation transformed into governance.

Section 7: Numbers, Measure, and the Abstraction of Reality

Once words and marks could preserve law, another form of translation began to dominate: number. Not just the scratch of tallies on bone, but the deliberate abstraction of the world into count, measure, and ratio.

Grain stored in jars could be tallied. Land boundaries could be measured with cords. Days could be marked in cycles, their intervals recorded in rows of symbols. Every abstraction was a leap: the messy, shifting fullness of life condensed into a clean mark that traveled further than memory.

Here translation ceased to be descriptive alone. It became predictive. To measure the floodwaters of the Nile was to read a future harvest. To calculate the movements of stars was to know when to sow. The signal was no longer only survival in the present; it was preparation for what had not yet arrived.

The mathematician was not yet a separate profession. The farmer, the builder, the priest—each performed calculations as a necessity. But as numbers accumulated into systems, as measures demanded consistency, a new kind of custodian emerged: the keeper of abstraction itself.

With numbers, the world could be re-rendered endlessly. A temple's stones were not only sacred—they were counted. A city's tribute was not only owed—it was measured. The human body was not only feared or revered—it was weighed. Every signal could be reduced to a quantity, and every quantity could be compared, traded, or commanded.

Translation had reached a new frontier: not fire, not stars, not marks, but pure relation— ratios, sums, proportions, laws. Mathematics was the most austere language yet, a tongue stripped of metaphor, yet filled with inexhaustible consequence.

Section 8: Geometry, Architecture, and the Shaping of Space

From number came shape. To count bricks was one act, to place them in precise alignment was another. Geometry emerged as the translation of abstraction into stone.

The pyramid was not merely a tomb— it was a proof in limestone. Each angle held ratios older than memory, ratios inherited from surveying fields, ratios first glimpsed in the shadow of a gnomon or the stretch of a rope across the earth.

Architecture became law carved into matter. The builder's cord was as binding as the priest's chant, the plumb line as commanding as the scribe's decree. Translation left the page and entered the skyline, making human order visible from miles away.

Cities rose as diagrams. Streets aligned with stars, walls traced defensive logic, temples reenacted cosmology in stone. Space itself became a script to be read, its proportions declaring: this is who we are, this is what we worship, this is what we demand.

To walk through a gate was to cross into meaning. To kneel beneath an arch was to inhabit proof. The body, once interpreter of fire and stars, was now surrounded by structures that spoke without words.

Geometry made translation inseparable from place. It was no longer only sound, mark, or number. It was stone that compelled obedience, design that declared order, space that silently instructed its inhabitants how to live, move, and believe.

Section 9: Instruments, Music, and the Harmonies of Order

If stone and geometry shaped space, music and number shaped time. A string stretched taut did not only produce sound— it revealed ratio. The division of length into halves and thirds was not decoration, but demonstration: harmony is structure made audible.

The drumbeat was more than noise. It was measure. It aligned bodies to a rhythm beyond any single will, binding many into one movement. Music translated number into pulse, abstraction into breath and gesture.

What seemed like joy or celebration was often calibration. Rowers on a river sang to synchronize their strokes. Armies marched to drums so their steps became one cadence. The temple echoed with chants that transformed individual voices into a single resonance.

Harmony became a metaphor for justice. Disorder in the city was like dissonance in a chord. A ruler's task was imagined as tuning— keeping the lyre of the people in balance. Philosophers would later describe the cosmos itself as a kind of music: the harmony of spheres, the inaudible order beneath all things.

Translation here was not only intellectual. It was visceral. The body felt it first, and the mind named it after. To live within rhythm was to accept that order could be heard, embodied, and repeated.

Music proved that translation was not only survival or record, but beauty. Signals could bind, but they could also elevate. The ratios that governed the stars could become the song that stirred the heart. What was once necessity had become something more: the discovery that structure itself could give rise to joy.

Section 10: Memory, Education, and the Lineage of Knowledge

Signals once lived in sensation alone. Then they were stored in marks, carried in ritual, and shaped into song. But survival demanded more than preservation. It demanded transmission.

So education arose— not in schools of stone at first, but in the discipline of repetition. The apprentice repeated the chant until his tongue obeyed. The child rehearsed the lineage of ancestors until each name was carried without error. The scribe copied the same line again and again, not as punishment, but as fidelity to the signal.

Memory became a craft. Techniques were invented: linking words to images, tying phrases to steps in a journey, fastening knowledge to melody. Each method was a scaffold, holding translation steady across the frailty of human recall.

To teach was not to invent. It was to transmit without distortion. Every elder's pride was not originality, but accuracy. Every pupil's duty was not to question, but to receive the signal intact. Education was the maintenance of translation, a lineage of carriers stretching across centuries.

Here responsibility grew heavy. To forget was more than failure— it was betrayal of the pattern that sustained life. The custodian who failed to teach risked more than shame; they risked the unraveling of the system itself.

In this way, memory hardened into institution. The

world was no longer translated only for the moment. It was translated for the future. Each repetition was a promise that what was once received would not be lost to silence.

Section 11: Power, Myth, and the Shaping of Destiny

To preserve a signal was one kind of strength. To guide it was another. Those who carried memory and law soon discovered they could also shape the horizon of belief.

The myth was not only a vessel of preservation; it was a compass of direction. A story told of origins was also a map of ends. If a people believed they descended from the sky, they would shape their rituals toward the heavens. If they believed they were born of the earth, they would return to the soil with reverence.

Translation thus began to mold destiny. What was once a neutral rendering of signals became a charged rendering of futures. The interpreter no longer simply relayed— they guided, persuaded, commanded. To carry the story was to carry the people, and to bend the story was to bend the arc of generations.

Myth armed leaders with more than memory. It

gave them legitimacy. A king was not merely powerful— he was chosen by gods, his throne aligned with the stars, his edicts extensions of cosmic law. Translation became weaponized, its signals sharpened into instruments of power.

The community was bound not only by survival, but by the vision of what they believed themselves to be. Rituals no longer only reminded— they declared. Symbols no longer only marked— they imposed. The interpreter had become an architect of fate.

Destiny itself was translated. What lay ahead was not discovered; it was narrated. And the narration, repeated long enough, took on the weight of inevitability. In this way, myth and power together turned translation into the architecture of collective futures.

Section 12: The Burden of the Interpreter

To be an interpreter was never neutral. The role carried weight, not only of memory and law, but of expectation, danger, and blame.

When the rains failed, it was the interpreter who had erred in reading the sky. When harvests withered, it was the custodian of ritual who bore suspicion. Translation was never free of risk— to speak for the signal was to answer for outcomes beyond human control.

Reverence and peril grew together. The same shaman who guided a tribe could be exiled when famine struck. The same priest who legitimized a king could be executed when omens soured. Authority rested on fragile ground: the thin line between trust in the signal and despair at its silence.

Yet the burden could not be abandoned. Without interpreters, the community unraveled. So each generation raised new custodians, teaching them not only the chants and symbols, but the resilience to carry both reverence and blame.

Translation thus became a vocation. It demanded discipline, memory, courage, and a willingness to be both elevated and condemned. The interpreter was never master of the signal, only its vessel. But the vessel was judged as if it commanded the current.

This paradox defined the role: to be responsible without control, to be accountable without authorship. The burden of the interpreter was the burden of carrying a system too vast to master, yet too vital to release.

Section 13: The Convergence of Signals

For centuries, signals had lived in fragments. Fire warned. Stars directed. Stone preserved. Ritual bound. Numbers measured. Each strand was powerful, yet partial.

In time, the strands began to weave together. Astronomy no longer stood apart from ritual— temples were aligned with solstices, altars became calendars, and sacrifices timed themselves to the heavens. Mathematics no longer belonged only to merchants— it shaped architecture, measured taxation, and refined the cycles of agriculture.

The convergence was not planned. It was demanded. Communities grew larger, cities denser, empires wider. Fragmented signals could not carry the weight. Translation had to become systemic.

Knowledge became layered. A law was inscribed in writing, timed by astronomy, enforced through ritual, justified by myth. The same event could now be read across many registers, each reinforcing the others, each binding the signal tighter into the fabric of life.

What had once been survival had now become civilization. Translation was no longer only about living through the moment. It was about coordinating multitudes, about weaving fragments into a network that could sustain generations.

In this convergence, the interpreter's role grew more complex. No longer guardian of a single signal, they were now mediators of many—balancing law with ritual, astronomy with agriculture, myth with mathematics. Translation had become a tapestry, and the interpreter, its fragile weaver.

Section 14: The Birth of Systems

Convergence demanded permanence. Signals woven loosely could fray. So civilizations began to harden translation into systems— stable forms that outlived their interpreters.

The library was one such system. Clay tablets stacked in rows, papyrus scrolls bound in chambers, stone inscriptions aligned in courtyards— these were not archives alone. They were attempts to make translation itself immortal. The signal no longer depended on fragile memory or shifting ritual. It lived in a structure designed to outlast flesh.

Law, too, became systematized. Not just decrees shouted in marketplaces, but codices etched in stone, articles ordered by number, clauses written to be copied and enforced. Law was no longer a ritual memory— it was a system of reference. A ruler could point not to a priest or elder, but to a text, and the text itself carried the command.

Education followed. What began as apprenticeship and rote repetition expanded into academies, schools of scribes, temples of instruction. Children were trained not only to repeat, but to inhabit the system of translation itself. The art of reading and writing was not freedom. It was initiation into a structured order that demanded fidelity to established symbols.

With these systems came new power. Translation ceased to be fragile and local. It became durable and centralized. Institutions emerged as custodians, claiming not only the duty to interpret, but the right to regulate who could interpret at all.

The birth of systems marked a threshold. Translation was no longer only the act of survival, but the architecture of civilization itself. It defined what could be remembered, what could be commanded, what could be

Section 15: The Invention of the Abstract

Once systems secured memory, a new freedom appeared: translation was no longer bound only to survival. It could now reach into realms where no harvest, no flood, no famine pressed against it.

Mathematics was the first to declare this independence. Numbers no longer measured only grain or stars. They began to measure themselves. Ratios of ratios, sums of sums— relations that existed in thought alone. Geometry extended beyond the surveyor's rope into proofs that lived nowhere but in the mind.

Philosophy soon followed. If myth translated signals into destiny, philosophy translated thought into argument. The same tools that once explained the season were turned inward, asking not only when to sow and when to reap, but what it means to know, what it means to live, what it means to be.

Abstraction was not an escape from translation— it was its refinement. The mind learned to render signals that came not from stars or stone, but from itself. The pattern of reasoning became a new text, and logic the grammar that held it together.

For the first time, translation moved faster than necessity. It no longer chased the urgency of survival. It generated new realms, realities of symbol and thought that could be debated, extended, and inherited like any other signal.

The invention of the abstract was the invention of possibility. To translate what had no immediate consequence was to declare that meaning itself could be pursued as a horizon. The human was no longer only a receiver of signals. It had become a constructor of conceptual worlds, even while still bound to the conditions that made this possible.

Section 16: The Fragility of Abstraction

Abstraction granted a new horizon, but with it came distance from the ground. What had once been anchored in fire, flood, and famine could now float free, spinning systems of thought unmoored from the signals that first compelled translation.

A proof could be elegant, yet useless to the farmer. A speculation could be dazzling, yet indifferent to hunger. The same tools that once preserved survival now generated worlds within worlds, realms where symbols multiplied without touching the soil that gave them birth.

This fragility was double-edged. On one side lay brilliance: concepts too vast for ritual or myth alone— infinity, harmony, justice, truth. On the other side lay peril: systems so abstract they could consume centuries, while fields lay untilled, while cities starved.

The interpreter's role became more precarious. No longer only a translator of sky and season, they were now negotiators between the concrete and the abstract. They had to decide which proof mattered, which philosophy could guide action, which speculation was worth binding into law.

Some succeeded, and civilizations flourished under their balance. Others failed, and their monuments to abstraction stood as ruins beside barren fields.

The fragility of abstraction was that it revealed humanity's new temptation: to treat translation as self-sufficient, as though meaning could survive without its roots in the world. It was a dangerous illusion, but also the very condition that made further discovery possible.

Section 17: The Standardization of Knowledge

Abstraction could scatter like sparks. Each proof, each speculation, each fragment of philosophy risked drifting into isolation. Civilizations answered this fragility by binding knowledge into standardized forms.

The canon was one such answer. Texts chosen, copied, and recited were declared authoritative. From a thousand voices, a few were set apart. From endless possibilities, a single lineage was defined. The canon stabilized translation, ensuring that every pupil began from the same signals, every interpreter worked from the same foundation.

The curriculum was another. Education ceased to be mere apprenticeship. It became structured progression. First letters, then words, first counting, then geometry, first myth, then philosophy. Knowledge was arranged into sequences, so the burden of inheritance was no longer random but orderly.

Standardization was preservation by narrowing. It ensured continuity, but at the cost of diversity. Voices outside the canon were silenced. Interpretations beyond the curriculum were dismissed. Translation, once the fluid art of responding to signals, was increasingly regulated by institutions that decided which signals mattered and which were to be ignored.

The scribe became examiner. The philosopher became schoolmaster. The interpreter, once a fragile vessel of survival, was now a gatekeeper of legitimacy.

Standardization stabilized civilization, but also hardened it. The system could now endure disruption, yet it also resisted innovation. Translation had become more durable— but also more rigid.

Section 18: The Silence of the Excluded

Every canon is a choice, and every choice is an exclusion. To declare one text sacred is to cast another into obscurity. To enshrine one philosophy is to leave countless others unheard.

The silence of the excluded was not empty. It was filled with songs that never entered temples, histories that were never carved in stone, mathematics that lived in trade routes and workshops but never in the scrolls of scholars.

Whole cultures found themselves erased. The victors' chronicles became "truth," while the defeated were reduced to rumor or myth. The canon was not only preservation— it was power deciding whose translation would survive.

Yet exclusion had its paradox. What was dismissed did not vanish. It survived in whispers, in oral traditions carried by those denied the script, in practices passed from mother to daughter, in tools perfected by craftsmen who never wrote a line. The excluded formed a shadow archive, an undercurrent of translation that official systems tried but failed to erase.

At times, these forgotten signals erupted back into history. A suppressed teaching resurfaced as heresy. A marginalized practice became a new science. A silenced people carried memory so long that even conquerors could not extinguish it.

The silence of the excluded was never absolute. It was the silence imposed by power, broken again and again by the persistence of signals that refused to vanish. Translation, no matter how systematized, could never fully control what the world continued to produce.

Section 19: The Return of the Forgotten

Suppression could delay, but it could not erase. Signals buried beneath conquest or neglect waited in silence, like seeds beneath ash, ready to germinate when conditions allowed.

A fragment of papyrus rediscovered in a desert ruin, a tablet unearthed by accident, a chant preserved in exile— each became a breach in the wall of official memory. Forgotten knowledge returned, sometimes as curiosity, sometimes as revolution.

The Greeks learned geometry from Egypt and Babylon, their "inventions" often restorations. Europe inherited astronomy and algebra from scholars of the Islamic Golden Age, who themselves had preserved and expanded the thought of Greece and India. Every renaissance was also a return, a translation of what had once been excluded.

The pattern was cyclical. What one age silenced, another recovered. What one empire buried, another revived. Translation was never linear. It advanced in spirals, retrieving what had been lost, binding it anew into fresh systems of meaning.

This return was never neutral. Knowledge re-entered changed. Signals carried through centuries of silence acquired new resonances. A myth reborn became scripture. A technique rediscovered became science. The forgotten did not merely return— it returned transformed.

Thus the history of translation was also a history of recurrence. What was once dismissed reappeared with power, reminding civilizations that no canon is final, no silence eternal, and no exclusion complete.

Section 20: The Cycle of Translation

Translation was never a straight line. It was a cycle, repeating under new names, shifting in scale but not in essence.

First came survival. Signals read in fire, stars, and flesh. Translation was urgent, direct, a practice of responding to conditions that could not be ignored.

Then came system. Marks carved, laws written, numbers standardized. Translation hardened into permanence, codified so that the signal might endure beyond a single voice or life.

Then came exclusion. Systems demanded boundaries. Some voices were admitted, others silenced. Translation became regulation, determining which signals were allowed to circulate and which were banished to obscurity.

And then, inevitably, came return. The forgotten resurfaced. Suppressed knowledge re-entered. Exiled voices found their way back into the center. The cycle renewed itself, reminding each civilization that no signal can be contained forever.

This cycle defined not only history but identity. Every human life mirrored it. Each child learned first in urgency, then in rules, then by rebelling against constraints, and finally by recovering what had once been discarded.

Translation was not an achievement to be secured. It was a rhythm that carried both civilizations and individuals alike. To survive was to enter it. To learn was to traverse it. To live was to be swept along its repetitions.

In recognizing the cycle, humanity glimpsed a deeper truth: that no translation is ever final. Every mark, every law, every canon, is provisional— a momentary resolution in an endless sequence. The signal always returns, changed but persistent, as relentless as life itself.

Section 21: Toward the Language of Proof

Each cycle demanded a stronger anchor. Stories could inspire, but they fractured in retelling. Rituals could bind, but they drifted with each performance. Writing could preserve, but even inscriptions carried the bias of those who carved them.

Civilizations longed for a translation that could resist decay, corruption, and dispute. They sought a language whose claims did not depend on the memory of elders, the power of priests, or the will of kings.

Mathematics appeared as that language. Not the crude tallies of grain and cattle, but the deliberate construction of systems in which truth could be demonstrated, step by step, with no appeal to authority beyond the logic itself.

Proof was the answer to fragility. It was translation distilled into necessity. A theorem once demonstrated did not require the charisma of a speaker or the sanction of an institution. It carried its own weight, compelling assent by structure alone.

The Pythagorean theorem would hold whether etched on papyrus, whispered in a school, or rediscovered on another continent. The proof could survive translation itself— moving across languages, cultures, and centuries without losing its force.

Here, at last, humanity discovered a translation that seemed immune to the cycle. Mathematics promised survival without distortion, system without tyranny, abstraction without collapse.

And yet, as the next chapter will show, this promise was both fulfilled and betrayed. For even in the purity of proof, the human interpreter remained. The cycle had not ended. It had only found its most austere expression.

Concluding Reflection

From firelight to starlight, from ritual chant to written law, from tally marks to theorems— every stage of human history was shaped by the act of translation.

We were never authors of the world. We were its readers. Signals struck us, and we rendered them into forms that could be carried, remembered, repeated, and obeyed.

Survival demanded urgency. Civilization demanded system. Power demanded exclusion. History demanded return. And through each cycle, translation adapted, expanded, and deepened, building the scaffolding of human life.

But the cycle was restless. No mark could hold forever, no ritual could bind without drift, no myth could remain unquestioned. Again and again, humanity searched for a translation that could resist decay.

That search led to mathematics. Numbers that measured harvests became numbers that measured proof. Geometry that surveyed fields became geometry that ordered the cosmos. Translation found its most durable form in symbols whose logic compelled assent regardless of culture, tongue, or time.

Chapter One has shown that the story of humanity is the story of signals carried across fragility. We inherited tongues not of our making, and we learned to survive by rendering them into meaning.

Now we turn to Chapter Two, to the language that promised permanence: the language of proof.

Chapter Two — The Language of Proof

— * —

Thesis

Mathematics standardized translation: symbols to law, procedure to result. Proof offered a language where meaning seemed immune to drift, a structure that could compel assent without ritual or decree.

Section 1: From Counting to Certainty

Counting was the beginning, but it was not enough. To know that there were twelve cattle or fifty jars of grain was useful for trade, yet it did not guarantee permanence. A tally could be misread, a memory disputed, a voice doubted.

Civilizations demanded something stronger— a way not just to record, but to compel agreement. This

demand gave rise to proof: a translation of intuition into necessity, a sequence of steps so clear that no interpreter could deny them.

The Egyptians measured their fields, the Babylonians tracked their skies, but it was in Greece that the desire for certainty was distilled into formal demonstration. Here, number ceased to be tally and became theorem. Here, geometry ceased to be surveying and became philosophy.

The proof was unlike any earlier signal. It did not beg belief. It required it. The right triangle carried within it a relation no ruler could alter, no priest could revoke, no ritual could amend. The Pythagorean theorem was not law imposed by humans. It was law discovered within structure itself.

Translation had reached a new register. For the first time, meaning could be stabilized without myth, without ritual, without the fragile trust in memory or authority. The signal seemed to speak for itself, and the human was only the recorder of its logic.

From counting to certainty, mathematics transformed the act of translation into something with the force of inevitability. It was no longer enough to say, "This is what happened." Now one could say, "This must be so, and could not be otherwise."

Section 2: Geometry as Demonstration

Proof did not remain only in words. It took shape in lines, circles, and constructions— marks on sand, scratches on wax, diagrams traced with compass and straightedge. Geometry became the theater of certainty.

To draw a square on the ground was to translate abstraction into visibility. To bisect an angle with a line was to show necessity with the hand. The figure was not illustration; it was argument embodied.

Here translation achieved a rare power: it bypassed persuasion. The eye could see what the voice declared. The mind was compelled not by tradition, not by fear, but by the clarity of relation. Once the diagram was drawn, disagreement dissolved. The signal carried its own authority.

Geometry turned space into logic. A temple could be beautiful, but a theorem was undeniable. The triangle inscribed in sand spoke with a permanence that rivaled the most enduring myth. It revealed that truth could be constructed step by step, so long as the interpreter obeyed the grammar of proof.

The role of the geometer was not invention, but demonstration. They were not authors of truth, but its arrangers. With each proof, translation grew more austere, stripped of ritual and myth, reduced to the bare necessities of structure.

Geometry taught humanity that reason could be made visible, that certainty could be enacted. A line, a circle, a square— simple forms, yet capable of carrying the weight of inevitability.

Section 3: Proof as Authority

Before proof, authority rested in bodies. The priest who spoke the ritual, the king who decreed the law, the elder who preserved the memory— each carried weight because of position, not because of inevitability.

Proof disrupted this balance. It carried an authority that no crown could match, an inevitability no ritual could overturn. A line drawn on sand and the reasoning that followed it could silence even the most powerful.

This was a revolution in translation. For the first time, truth was not dependent on hierarchy. It was accessible to anyone who could follow the steps. The logic compelled assent without appeal to rank or role.

This did not erase power— kings still ruled, priests still chanted, elders still guided— but it introduced a rival, a standard against which their words could be tested. Where once dissent was heresy, now it could be correction. Where once obedience was demanded, now reasoning could be demanded in return.

In this way, proof democratized and destabilized. It offered a path of authority that belonged not to birthright, but to clarity. The interpreter of proof did not command belief through fear or inheritance, but through demonstration itself.

Authority shifted. The temple and throne remained, but beside them stood the diagram and the theorem. And while kings and priests could command bodies, proof commanded minds.

Section 4: The Discipline of Logic

Geometry made certainty visible, but its reach was limited to shapes and measures. The next step was more radical: to treat thought itself as something that could be ordered, demonstrated, and tested.

Thus arose logic— not as rhetoric, not as persuasion, but as structure. Aristotle's syllogisms, the rules of inference, the catalog of fallacies—these were attempts to render thought as disciplined as number and line.

Where proof once applied only to triangles and circles, it now applied to arguments and ideas. "Every human is mortal. Socrates is human. Therefore, Socrates is mortal." Here was no myth, no decree, only necessity. The conclusion carried the same inevitability as a theorem inscribed in sand.

Logic was translation stripped to its core. It revealed that the mind, like the world, could be compelled into coherence. No priest or king could alter the consequence of a valid syllogism. The authority of proof had entered the very machinery of thought.

With logic, interpretation itself was standardized. To reason well was no longer talent or charisma—it was obedience to structure. The interpreter became less a performer of wisdom and more a technician of thought.

The discipline of logic carried immense promise. It offered a method to resolve disputes without violence, to pursue truth without reliance on tradition, to build knowledge on a foundation that did not shift with culture or power. It was the audacious claim that clarity could be engineered.

Section 5: The Expansion of Proof

Once the discipline of logic was established, the ambition of proof could not remain contained. It began to move outward, stretching from triangles and syllogisms to the heavens, to motion, to matter itself.

Astronomy was among the first to be reshaped. The stars had always been signals— guides for planting, omens for ritual. But now they were subjected to geometry and number. Their cycles could be calculated, their motions predicted. The sky was no longer mythic alone; it became demonstrable.

Mechanics followed. Weights and pulleys, levers and wheels— once tools of craft— were brought into the domain of reasoning. Principles of balance and motion were rendered in diagrams, their behavior derived step by step. The craftsman's art became the geometer's proof.

The natural world itself was redrawn as something that could be reasoned with. Plants, animals, bodies— all invited analysis, their patterns translated into order. What had once been the domain of myth or magic became subject to the language of demonstration.

This expansion was intoxicating. Proof promised not only stability of thought, but mastery of the world. To explain the heavens with geometry was to glimpse control of navigation. To explain machines with logic was to glimpse control of labor. To explain matter with number was to glimpse control of nature itself.

The expansion of proof was also a transformation of ambition. What began as the quiet discipline of necessity became a project of conquest— a way of translating the entire cosmos into relations that could be derived, repeated, and commanded. The language of proof had left the tablet and the classroom. It was now inscribing itself onto the very structure of reality.

Section 6: Proof and the Promise of Universality

The more proof succeeded, the more it seemed to offer not just clarity, but universality. A theorem drawn in Greece could be rediscovered in India. A logical syllogism spoken in one tongue could be carried into another without distortion. Proof appeared to transcend the accidents of culture, the fragility of memory, the authority of kings.

This promise was intoxicating. Here, at last, was a language that did not splinter. While myths fractured into countless variations, while rituals drifted with each generation, proof retained its force. No matter who invoked it, a valid demonstration compelled assent.

The claim was audacious: that beneath all diversity, there existed a single grammar of necessity. The triangle's angles summed to a straight line whether in Athens or Babylon, Alexandria or Chang'an. Here was a signal not of one culture, but of structure itself.

The interpreter's role seemed diminished. The proof did not depend on charisma or memory. It lived in the steps themselves. Anyone who could follow the reasoning could arrive at the same conclusion. For the first time, translation appeared to have no author at all— only participants in a process larger than themselves.

This universality reshaped ambition. If proof could stabilize geometry, could it not stabilize politics? If logic could order arguments, could it not order society? Dreams of universal law, universal ethics, universal systems of thought all sprang from the conviction that proof was not just a tool, but a foundation for everything.

The promise of universality was proof's greatest gift— and its greatest temptation. For while structure compelled agreement, human life was never only structure. The danger of proof lay not in its limits, but in forgetting that it had them.

Section 7: The Limits of Proof

For all its clarity, proof was never absolute. Its universality was powerful, but only within the boundaries it defined for itself. Outside those boundaries, life remained unruly, resistant to neat demonstration.

A triangle could be proven, but justice could not. A ratio could be derived, but desire defied calculation. Logic compelled agreement in syllogisms, but human motives wandered beyond necessity.

Even within mathematics, limits appeared. Paradoxes haunted the Greeks: Zeno's endless divisions, the incommensurability of diagonal and side, the unsettling discovery of irrationals. Proof could stabilize much, but it also revealed fissures— truths that strained against its own methods.

The interpreter of proof learned humility. The steps compelled assent, but only after assumptions were chosen. Choose different axioms, and the world rearranged itself. The certainty of proof was not the certainty of nature, but the certainty of structure built on foundations.

Still, proof held allure. Its limits did not diminish its power. They only made clearer what it was: not a universal solvent of all questions, but a disciplined language capable of holding necessity within the domains it could reach.

The dream of universality survived, but it was tempered. Proof was no longer imagined as the key to all things, but as one key among many. Its strength lay in its boundaries, its force in its discipline. Beyond it, the human still wandered—but within it, the human had found a rare and precious stability.

Section 8: Proof as Inheritance

Proof was never only discovery. It was also inheritance. Each demonstration, once established, became a seed carried forward, a signal preserved for future minds to receive.

The Euclidean tradition exemplified this. Propositions ordered into a system, each theorem resting on the last, each proof a brick in a structure designed to endure. It was more than geometry. It was pedagogy, a curriculum of certainty that could be copied, translated, and taught.

In Alexandria, scribes reproduced these steps. In Baghdad, scholars translated them into Arabic. In Córdoba, they were copied again into Latin. Across continents and centuries, the same diagrams reappeared, the same arguments compelled assent. The proof became a lineage, its vitality ensured by its portability.

Unlike myth, which fractured, or law, which demanded enforcement, proof traveled lightly. It did not require loyalty to a king or devotion to a god. It required only patience, a surface to write upon, and a mind willing to follow the steps.

In this way, proof shaped education. To be schooled was to inherit demonstrations older than the student, older even than the teacher. Each pupil rehearsed not the charisma of a master, but the clarity of relations that compelled obedience across generations.

Proof thus became a cultural backbone. It offered civilizations not just tools, but continuity— a chain of reasoning that could survive the collapse of empires, the shifting of languages, the rise and fall of rulers. To inherit proof was to belong to a lineage that exceeded borders and bloodlines.

Section 9: Proof and the Discipline of Doubt

Proof was not only a language of certainty. It was also a training in doubt. For to demonstrate one claim was to expose all others to scrutiny. Every proof declared: what is not shown cannot yet be trusted.

This discipline sharpened the mind. It forced interpreters to ask: what assumptions am I making? what steps can be justified? where does necessity begin, and where does it collapse into error?

Thus proof cultivated a vigilance that myth and ritual had never required. It taught a posture of questioning, not to undermine truth, but to stabilize it. The same structures that compelled assent also compelled interrogation.

Over time, this discipline spread beyond mathematics. Philosophers adopted it, testing arguments for coherence. Natural thinkers adopted it, examining whether claims about plants, stars, and bodies could be derived rather than assumed. The culture of proof became a culture of testing, and testing bred progress.

Doubt was no longer a threat to authority— it was a requirement of authority. The interpreter gained legitimacy not by suppressing questions, but by anticipating them, answering them with steps too clear to deny.

In this way, proof reshaped not only what was known, but how knowing itself was practiced. It turned curiosity into method, skepticism into discipline, and disagreement into refinement. The gift of proof was not certainty alone. It was the courage to demand that certainty be earned.

Section 10: Proof and the Architecture of Knowledge

Each proof was a stone, but stones alone do not make a city. Over centuries, demonstrations were stacked, aligned, and interlocked into structures vast enough to house generations of thought.

Euclid's Elements was one such architecture. Not merely a collection of theorems, but a cathedral of reasoning. Axioms formed its foundations, definitions its columns, propositions its arches, corollaries its windows of light. It was a building in which the mind could dwell, moving from chamber to chamber, each space designed with precision.

Other traditions raised their own architectures. Indian mathematicians constructed systems of algebra. Chinese scholars developed structures of number, harmonics, and astronomy. Islamic thinkers carried these architectures forward, translating, expanding, refining, linking ancient bricks to new foundations.

The architecture of proof was more than accumulation. It was coherence. Each new result was not added loosely, but fitted into place, tested for alignment with what stood before. The system grew not as a heap, but as a structure, each stone reinforcing the next.

This architecture gave civilizations a new kind of durability. Even when empires fell, their intellectual cathedrals endured. Copied and carried, they rose again in new lands, rebuilt stone by stone from memory and manuscript.

Proof had become not just a method, but an architecture— a way of building entire worlds of thought that could be inherited, expanded, and lived within. The human mind had found a dwelling not in myth, not in ritual, but in the geometry of necessity.

Section 11: Proof and the Training of the Mind

The architecture of proof was never only shelter. It was also a school. To move through its corridors was to train perception, patience, and precision. Each proposition demanded more than memorization. It demanded the ability to follow necessity step by careful step.

Students of mathematics were not taught only results. They were taught a way of thinking. The copying of diagrams, the rehearsal of demonstrations, the repetition of logic— all served to engrain a discipline that shaped minds as much as it preserved knowledge.

Proof trained the eye to seek relation. It trained the ear to detect inconsistency. It trained the hand to construct with accuracy. It trained the tongue to argue with restraint. It was less a collection of facts than an apprenticeship in clarity.

This training extended beyond mathematics. Rhetoric was reshaped by logic. Philosophy was disciplined by demonstration. Even law borrowed from the rigor of proof, demanding evidence and procedure instead of mere decree.

The interpreter who had once relied on charisma was now expected to master structure. The mark of wisdom was not only what one remembered, but how one reasoned. Proof became a measure of mind— a way of distinguishing the careful from the careless, the disciplined from the impulsive.

Thus, proof functioned both as archive and as exercise. It stored truths across centuries, and it trained each new generation to carry them forward with precision. The true inheritance of proof was not only the theorems themselves, but the habits of thought that proof instilled in those who studied it.

Section 12: Proof and the Expansion of Inquiry

The discipline of proof did not confine itself to geometry and logic. Once minds were trained to demand clarity, they carried that posture into other realms. The method spread like a contagion, seeking order wherever signals seemed chaotic.

Astronomy absorbed it first. The sky was no longer only an omen. It became a field of predictable relations. Positions of planets were measured, cycles of eclipses cataloged, motions of stars described in geometric models. The heavens themselves were pulled into the architecture of proof.

Mechanics followed. What builders and craftsmen

had known in practice was now subjected to demonstration. Levers and pulleys were no longer only tools— they became principles, derived from axioms, framed as laws. The work of the artisan became the theorem of the philosopher.

Medicine, too, began to shift. Observation of symptoms was not enough. Classification, reasoning, and hypothesis entered the healer's practice. Though still fragile, it was the first step toward translating the body into a system governed by principles rather than incantations or guesswork.

Each expansion carried risk. Not every domain yielded easily to proof. The body, the weather, the mind— these resisted clean demonstration. But even where certainty faltered, the ambition remained: that the world could be ordered, that signals could be stabilized, that reality itself could be translated into necessity.

Proof had thus become more than method. It was a worldview. To think was to demonstrate, to know was to prove, to trust was to demand necessity. Inquiry, once scattered, was being pulled into alignment under the architecture of proof.

Section 13: Proof as Cultural Prestige

As proof expanded across domains, it began to carry weight beyond its immediate use. To master demonstration was not only to reason well— it was to belong to a culture of refinement, a tradition of intellect that elevated its practitioners above mere survival.

In Athens, to argue with clarity was to be counted among the educated. To trace diagrams in the sand was to signal membership in a lineage that valued necessity over persuasion, logic over charisma, structure over story.

Elsewhere, too, proof became a marker of status. In India, mathematical schools were honored for their precision and subtlety. In the Islamic Golden Age, to preserve and extend Greek demonstrations was a badge of civilization itself. In Europe, centuries later, the recovery of proofs would be celebrated as the rebirth of reason.

Proof was thus more than a tool. It was prestige, a cultural ornament as well as an intellectual discipline. The ability to demonstrate a theorem or refute an argument with logic conferred respect that no wealth or weapon could guarantee.

The geometer, the logician, the mathematician—

these figures were elevated, not because they wielded power directly, but because they seemed to hold a language immune to corruption. To be skilled in proof was to stand close to the eternal, to speak in the grammar of necessity rather than the noise of opinion.

Civilizations used proof as a mirror of themselves. To display mastery of demonstration was to declare sophistication. To preserve a corpus of proofs was to claim continuity with the wise. Proof became not only an intellectual practice but a civilizational boast: the evidence that a people had risen beyond mere survival into the discipline of truth.

Section 14: Proof and the Temptation of Control

Prestige is never idle. Once proof became a badge of refinement, it was inevitable that rulers, priests, and thinkers alike would seek to wield it as an instrument of control.

If proof could compel assent in mathematics, why not in politics? If logic could order arguments, why not order society? If demonstration could silence dissent in geometry, why not silence it in law?

This temptation reshaped ambition. Constitutions were framed as if they were theorems. Religious doctrines were defended with syllogisms instead of revelation alone. Economic theories were presented with the rigor of geometry, as though wealth and labor were triangles awaiting proof.

The allure was clear: proof seemed incorruptible. Unlike myth, it did not fracture. Unlike ritual, it did not drift. Unlike law, it did not require enforcement by threat. Its necessity appeared self-sustaining.

But the temptation was also dangerous. To treat human life as geometry was to mistake living complexity for fixed structure. To demand certainty in matters of justice or love was to impose rigidity where fluidity was required. The desire for proof in all things risked reducing the richness of experience to a brittle system of false necessity.

Still, the dream persisted. Philosophers and rulers alike imagined cities built like proofs, societies aligned like diagrams, human lives arranged with the clarity of axioms. It was a vision both seductive and impossible, born of the same brilliance that had once drawn triangles in sand.

Section 15: Proof Under Challenge

For all its prestige, proof could not escape the pressure of reality. The same rigor that made it compelling also exposed its fragility.

Skeptics emerged early. Sophists in Greece questioned whether demonstrations truly revealed necessity or only concealed assumptions. If a proof depended on axioms, what gave those axioms their authority? Was certainty absolute, or only inherited from unprovable beginnings?

Paradoxes deepened the doubt. Zeno divided motion into infinite steps and left philosophers stumbling. Irrational numbers shattered the harmony of ratios, revealing lengths that could not be expressed by any neat fraction. Even within the architecture of geometry, cracks appeared.

Later, as inquiry expanded, the challenge grew sharper. Observation sometimes refused to align with inherited demonstrations. The motions of planets strained against ancient geometric models. The fall of bodies did not follow the proofs once assumed. Nature itself resisted the elegance of human diagrams.

Proof was forced into confrontation with the world. It was no longer enough to compel minds with logic and diagram. The demonstration had to face experiment, the theorem had to meet evidence.

This was not the end of proof, but its transformation. What had once lived in sand and parchment would now have to survive in falling apples, turning planets, and moving machines. The authority of proof was being tested— and, in that testing, prepared for rebirth.

Section 16: Proof and Experiment

The challenge of reality did not destroy proof. It forced it to adapt. Where demonstration alone could no longer suffice, it joined hands with observation. A new partnership was born: the marriage of reasoning and experiment.

Galileo dropped bodies from towers and rolled spheres down inclined planes. The results were not left to intuition, but measured, timed, and compared. Geometry still structured the explanation, but now the world itself offered the evidence that confirmed or refuted the claim.

This union marked a turning point. No longer could diagrams rule untested. Every theorem that claimed to describe nature was now accountable to experience. Proof alone compelled the mind, but proof with experiment compelled the world.

The scientific method emerged from this convergence. Hypothesis, deduction, test, revision— a cycle where reasoning and evidence were bound together in mutual discipline. Mathematics became the scaffolding, but experiment became the foundation.

This synthesis expanded the reach of inquiry. The motions of planets were no longer mythic, but calculable. The fall of bodies was no longer mystery, but law. Light, sound, pressure, heat— all entered the field of demonstration tempered by observation.

The interpreter's role changed again. They were no longer only a builder of arguments, but a tester of reality. The chalkboard and the laboratory became twin arenas of truth. In their union, knowledge acquired a durability that neither alone could have secured.

Section 17: Proof in the Age of Science

When proof joined experiment, a new age began. What had once been the domain of philosophers became the vocation of scientists. The old dream of necessity was now sharpened by measurement, anchored not only in reason but in the stubborn testimony of matter itself.

Kepler reshaped the heavens with ellipses. Galileo charted the acceleration of falling bodies. Newton united earth and sky with a law of universal gravitation, a single relation proved in mathematics and confirmed in observation. Here was proof not only of triangles, but of planets, tides, and comets.

The Age of Science was the age of translation reborn. The signals of nature— once read in myth, ritual, and story— were now rendered in equations. The arc of a cannonball, the orbit of a moon, the sway of a pendulum— all became sentences in the new grammar of proof.

The authority of kings and priests waned as the authority of experiment and demonstration grew. A law of motion compelled assent more firmly than a royal decree. A principle of optics carried weight that no chant could rival. The balance of power shifted, not through rebellion alone, but through the quiet inevitability of truths that no hierarchy could deny.

Yet this authority was fragile in another way. For the first time, truth was expected to evolve. Proof in science was never final. Each law held until a better one arrived. Certainty became provisional, anchored in what was tested and repeatable, but always awaiting revision.

Thus the Age of Science revealed both the triumph and humility of proof. It could bind the heavens to the earth, yet it carried within itself the duty to be tested, doubted, and replaced. Proof had become not the end of inquiry, but its engine.

Section 18: The Democratization of Proof

Once confined to temples, academies, and courts, proof began to escape its walls. The printing press multiplied diagrams and arguments. Books once chained to libraries spread into the hands of merchants, artisans, and students. The logic of demonstration was no longer a secret— it was a public inheritance.

Pamphlets carried geometrical puzzles into taverns. Almanacs charted eclipses for farmers. Sermons borrowed syllogisms to persuade. Proof was no longer only a tool of philosophers— it was entertainment, persuasion, and education for the many.

Schools adopted its rigor. Children copied figures, memorized axioms, and rehearsed demonstrations not to become mathematicians, but to be disciplined by the clarity of reasoning. The discipline of proof became a social expectation, a mark of literacy alongside reading and writing.

The democratization of proof reshaped the boundaries of authority. The artisan who could calculate, the sailor who could navigate with geometry, the farmer who could read astronomical tables— all became empowered interpreters of signals once reserved for elites.

This diffusion was disruptive. Power once monopolized by priests and scholars was now scattered across society. Translation was no longer only inherited from the top— it was practiced in households, workshops, and markets.

The proof thus became not only cultural prestige, but civic expectation. To belong to a modern society was to be capable of following a demonstration, to accept that truth was not decreed, but shown. The architecture of knowledge had become a public dwelling.

Section 19: Proof and Revolution

The democratization of proof carried consequences that no philosopher or monarch could fully control. Once ordinary people learned that necessity could be demonstrated, they began to demand the same rigor from the structures that governed their lives.

If triangles required justification, why not taxes? If axioms could be questioned, why not decrees? If every step in a theorem had to follow with clarity, why should laws or policies be accepted without demonstration?

Thus the discipline of proof became the seed of revolution. The same reasoning that stabilized geometry began to destabilize kingdoms. The Enlightenment was not only an age of science— it was an age of argument, where citizens demanded from rulers the same necessity they had been taught in schoolrooms.

Declarations of independence read like syllogisms. Rights were derived as if they were theorems. Constitutions were drafted as if they were systems of axioms, from which freedoms and obligations could be logically deduced. Revolutions carried muskets, but they also carried proofs— arguments written in the ink of necessity.

Yet the translation was imperfect. Societies were not triangles. People were not axioms. The attempt to govern humanity with the same rigor as mathematics revealed both brilliance and brutality. Certainty inspired liberation, but rigidity birthed terror.

Still, the imprint of proof remained. It taught generations to expect clarity, to resist arbitrary command, to treat persuasion not as sufficient, but as accountable to reasoning. The architecture of demonstration had escaped mathematics, and now pressed against the foundations of society itself.

Section 20: Proof and the Machinery of Progress

Revolutions shook thrones, but proof did not stop at politics. Its discipline seeped into the workshop and the forge, transforming craft into industry.

Where once machines were built by trial and error, they were now designed by principle. The lever and pulley became equations. The steam engine was refined by calculations of pressure, volume, and efficiency. Bridges rose from rivers not by instinct, but by geometry and stress analysis. Translation, once survival and law, was now steel and smoke.

The Industrial Revolution was also a revolution of proof. Blueprints replaced guesswork. Standards replaced improvisation. The architecture of demonstration became the architecture of machines, each bolt and beam positioned by derivation.

Progress acquired a new rhythm. Inventions could be scaled, replicated across continents, because they rested not on craft alone but on principles that traveled as easily as proofs. What was once the secret of one workshop became the inheritance of all who could read diagrams and follow equations.

This machinery of progress reshaped power. Empires no longer relied solely on armies or gold. They relied on the capacity to translate proof into production, to harness necessity for speed, scale, and control.

Yet even here, proof revealed paradox. The same rigor that enabled progress also demanded conformity. Machines obeyed equations without error, but humans did not. To live in the age of proof-made-machinery was to experience both liberation and constraint— a world expanding with possibility, yet increasingly bound by the systems that proof itself had built.

Section 21: The Abstraction of the Invisible

Proof had conquered stone, sky, and machine. Its next frontier was the unseen. Human senses reached their limit, but reasoning extended further, building structures for phenomena that could not be touched, smelled, or watched with the naked eye.

Electricity crackled in storms and sparks, yet its

laws were invisible. Proof gave them form. Equations described charges and currents, fields and forces, long before wires carried power across continents.

Magnetism pulled compass needles, but its essence was hidden. Geometry and algebra revealed that it followed patterns as strict as triangles and levers. Invisible lines of force became diagrams on parchment, turning mystery into law.

Atoms, too, were imagined through proof before they were glimpsed in instruments. Parmenides guessed, Democritus speculated, but mathematics tested, refined, and confirmed. Equations described their weights, their motions, their bonds— ghostly structures derived by necessity, long before microscopes made them visible.

This was proof's boldest leap. It dared to translate what the body could not sense. The visible world was revealed as only a surface of deeper structures, governed by relations abstract but inexorable.

Here, proof achieved a strange paradox. It made the invisible more certain than the visible. A falling apple could deceive, but the equation of gravity compelled. A flash of lightning could terrify, but Maxwell's laws endured. The eye faltered, yet the demonstration held.

The abstraction of the invisible marked proof's transformation into modern science. No longer confined to what was seen, it claimed authority over what must exist— even when no eye could yet confirm it.

Section 22: Proof and the Fragmentation of Certainty

The deeper proof pressed into the invisible, the more it uncovered cracks in the smooth surface of certainty.

Infinity was one such fissure. Calculus revealed that motion and change could be tamed by infinitesimal steps, yet infinity itself defied final containment. Each proof that wielded it carried a shadow of unease: was this necessity, or an illusion of precision stretched too far?

Probability was another. Once certainty's rival, it entered the architecture of reasoning itself. Proof no longer spoke only of what must be, but of what was likely. The language of necessity was fractured by the language of chance.

Quantum physics deepened the fracture. Where once particles obeyed strict trajectories, now they shimmered in probabilities, their behavior demonstrable only in distributions. Proof survived, but it spoke in statistics rather than absolutes, in wave functions rather than diagrams.

Even mathematics revealed its own limits. Gödel proved that within any formal system rich enough to hold arithmetic, there would be truths that could not be proven. The dream of a complete architecture of certainty collapsed under the weight of proof itself.

What had once promised universality was now fragmented. Certainty endured in places, but not everywhere. The authority of proof remained powerful, yet no longer absolute.

The fragmentation of certainty did not end the story of proof. It only clarified what it had always been: a discipline of structure, bounded by its own assumptions, capable of astonishing reach, yet forever aware of its horizons.

Section 23: Proof in the Age of Machines

For millennia, proof was a human labor. Lines drawn by hand, syllogisms spoken aloud, equations scratched on parchment. The interpreter was flesh, and the demonstration lived in voice, eye, and hand.

But with machines came a new horizon. Computation transformed proof from an act of demonstration into an act of verification. Algorithms traced steps faster than any human, testing relations too vast for memory, too intricate for chalkboards.

What Euclid arranged with compass and straightedge, computers extended into millions of cases. The four-color theorem, once resisted for lack of elegance, was proven with the aid of exhaustive calculation— a proof no single person could hold, but which a machine could confirm.

This was both triumph and crisis. Triumph, because new truths emerged, truths unreachable by human patience alone. Crisis, because the role of the interpreter shifted. The human no longer followed every step. They trusted the machine's traversal of logic, a black box of necessity that mimicked the rigor once carried by the geometer's hand.

Proof in the age of machines was no longer a performance of clarity. It became an architecture of trust— trust in algorithms, trust in silicon, trust that the translation of reasoning into code was faithful and complete.

The age of machines revealed proof's newest paradox: that the very discipline once designed to liberate truth from authority now demanded new forms of authority— the authority of computation, the authority of systems too vast to be seen whole. Proof had left the sand and parchment. It now pulsed through circuits and code.

Section 24: Proof and Artificial Intelligence

If computation transformed proof into verification, artificial intelligence pushed further. Machines no longer only checked steps. They began to propose them.

Automated theorem provers generated demonstrations beyond the reach of individual minds. Neural networks explored spaces of conjecture and suggested patterns no geometer had drawn. The line between human reasoning and machine reasoning blurred.

Here the old question returned with new force: what is proof without understanding? If a machine produces a sequence of steps that no person can follow in full, does necessity still compel us? Is truth a matter of comprehension, or of structure alone?

For some, the answer was liberating. The universe had never promised that human minds would be its sole interpreters. If machines could extend the discipline of proof, then truth itself was being widened, welcoming new translators into the lineage.

For others, the answer was unsettling. Proof had always been performance as well as result. The beauty of a theorem was in its elegance, its clarity, its ability to illuminate. What was gained if truth became unreachable, locked inside the silent efficiency of code?

Artificial intelligence revealed a deeper shift. Proof was no longer bound to human scale. It became planetary, computational, distributed— a signal carried not only in minds, but in servers, algorithms, and networks. The interpreter was no longer a single figure, but a collaboration between flesh and machine.

Here, the ancient lineage of proof met the newest frontier of translation. The circle drawn in sand now echoed in circuits of silicon. The human was still present, but no longer alone.

Section 25: Proof and the Question of Trust

Proof had once been the antidote to trust. Where myth demanded belief, proof demanded only attention. Its steps compelled assent without appeal to authority. To follow the reasoning was to be convinced.

But in the age of machines, trust returned through the back door. The interpreter could no longer follow every step. The demonstration sprawled across millions of calculations, traced in silicon too fast and too vast for the human eye. The conclusion was clear, but the path was hidden.

This inversion was unsettling. The discipline designed to free truth from hierarchy now required faith in algorithms, in programmers, in systems opaque to most of those who invoked them. Proof became less performance, more infrastructure— a thing maintained rather than witnessed.

Some resisted. They longed for proofs elegant enough to be held in the mind, proofs that restored the intimacy of reasoning. Others embraced the shift. They argued that necessity was never about comprehension, but about structure. If the structure held, then proof endured, whether or not it could be read line by line.

This debate revealed proof's new vulnerability. It was no longer simply a matter of logic. It was a matter of trust in the systems that carried logic beyond human grasp.

The question of trust would not vanish. It shadowed every new theorem verified by machine, every conjecture settled by algorithm. To believe in proof was once to believe in necessity. Now it was also to believe in the fidelity of machines.

Section 26: Proof at the Edge of Comprehension

The dream of proof was clarity. A line drawn in sand, a syllogism spoken aloud— every step visible, every conclusion transparent.

But modern proofs began to swell beyond the reach of that dream. Some sprawled across thousands of pages. Others filled libraries with cases, each verified by machine, but unreadable by any single mind. The demonstration survived, yet the intimacy was gone.

Mathematics at its frontiers began to resemble cartography of the infinite. Teams of thinkers collaborated across decades, each mastering only a fragment, trusting the rest to colleagues and machines. The proof became a collective event, a cathedral raised by many hands, its design understood in outline but opaque in detail.

This edge of comprehension redefined what proof meant. No longer the solitary act of a geometer kneeling in sand, it became a networked endeavor, part argument, part verification, part trust in systems that no one could survey entire.

For some, this was liberation. Humanity had always been limited. Now proof extended beyond those limits, a demonstration of what collaboration could achieve. For others, it was disquieting. If no single mind could hold the whole, was proof still proof, or had it become something else— a ritual of confidence in structures too vast to grasp?

At the edge of comprehension, proof revealed its dual nature. It was still necessity, but necessity refracted through scale. Clarity remained, but it was clarity distributed across minds and machines, a light no longer held in one hand, but scattered across the horizon of the collective.

Section 27: Proof and the Horizon of the Unknown

At the farthest edges of inquiry, proof encountered questions that refused to resolve. Not for lack of brilliance, nor for lack of effort, but because the structure of proof itself trembled at their boundaries.

The continuum hypothesis lingered unresolved— a claim about infinity neither provable nor disprovable within accepted axioms. It revealed that certainty could fracture, that truth might depend on which foundations one chose to begin with.

The problem of P versus NP carried similar weight. It asked whether every solution quickly checked could also be quickly found. No demonstration yet answered it, yet the question haunted computation, economics, biology, cryptography. Entire industries relied on the silence of proof, on the abyss between what could be verified and what could be derived.

These horizons unsettled the dream of necessity. Proof was supposed to compel agreement. But here were problems where agreement could not be reached, where the very structure of reasoning seemed incomplete.

For some, this was despair. If proof could not settle the deepest questions, what remained of its promise? For others, it was revelation. The horizon of the unknown was not a failure of proof, but a reminder of its nature: a method bounded by assumptions, powerful within its walls, yet honest enough to reveal where its walls stood.

Thus proof at its horizon did not collapse into silence. It became a mirror of humility. It taught that even necessity had edges beyond which the human and the machine alike must wait.

Section 28: Proof as a Human Mirror

Proof was never only about triangles, equations, or logic. It was also about us. In every demonstration, humanity saw its own reflection— the desire for clarity, the discipline of patience, the humility of limits.

When a theorem compelled assent, it revealed more than necessity. It revealed the strange capacity of the mind to align with structures beyond itself, to step into patterns that existed long before human eyes and would endure long after.

When a proof collapsed under paradox, it revealed more than fragility. It revealed the boundary of our reach, the edge where language falters, and assumptions betray themselves. Our failures in proof were as telling as our triumphs— they displayed the contours of our limits, drawn with the same sharpness as our successes.

Proof was aspiration embodied. Each demonstration carried with it the longing for permanence, the hope that something of our thought might resist decay. To prove was to seek a voice that no time, culture, or power could silence.

And yet proof was also confession. Every axiom declared: here we begin, though we cannot say why. Every limit acknowledged: here we stop, though we wish to go further. The mirror of proof reflected both our strength and our surrender.

To study proof, to inherit it, to extend it— was therefore not only to know mathematics. It was to know ourselves, to glimpse in necessity the restless condition of being human: always seeking clarity, always facing boundaries, always translating signals into the fragile architecture of understanding.

Section 29: The Future of Proof

Proof has always been restless. It began in sand and stone, traveled through parchment and press, and now pulses through circuits of silicon. Its future will not be still.

Artificial intelligence will not only verify but generate new demonstrations, charting regions of mathematics no human eye may ever fully grasp. Distributed networks will host proofs not in single manuscripts but in living systems, updated and refined in real time. Necessity itself may become a moving current, flowing across machines and minds.

Visual languages may rise beside symbols, proofs

embodied in simulations, animations, and models, where structure is seen, felt, and tested dynamically. The classroom of the future may not copy diagrams onto chalkboards, but enter proofs as environments to be explored.

Yet the question of trust will remain. Who will verify what no single mind can hold? What will count as understanding when proofs live in algorithms too complex for human inspection? The mirror of proof may tilt from clarity to collaboration, from solitary reasoning to networks of shared belief.

And perhaps beyond proof as we know it, new forms of necessity will appear. Biological computation, quantum reasoning, languages not yet imagined— each may expand what it means to compel assent, to stabilize translation, to glimpse the architecture of truth.

The future of proof will be, as always, the future of humanity itself. Our methods will change, our tools will evolve, but the longing will remain the same: to touch necessity, to translate the signals of the world into steps that no power can deny, to build within uncertainty a dwelling of permanence.

Concluding Reflection

From the first marks in sand to the circuits of machines, proof has been humanity's most ambitious translation. It sought not only to preserve signals, but to compel assent, to bind minds across distance, time, and culture.

It promised certainty, and often delivered. It disciplined doubt, and trained generations in clarity. It built architectures of knowledge that endured beyond empires. It fueled revolutions, powered machines, and opened windows into the invisible.

And yet proof also revealed its limits. It fractured under paradox, strained at infinity, trembled before probability, and confessed its boundaries in incompleteness. Even in its triumphs, it mirrored the fragility of its interpreters— finite beings seeking permanence in an infinite world.

Now, in the age of machines and networks, proof is less a solitary act than a collective horizon. Its authority rests not only in clarity, but in trust. Its future belongs not only to mathematics, but to the wider systems of reasoning that will emerge from our union with computation.

Yet the essence remains. To prove is still to declare that something must be so. It is still the human longing to glimpse necessity amid flux, to stand, if only for a moment, on ground that will not shift.

Thus Chapter Two closes where it began: with proof as the most disciplined language of translation, born from counting, refined in geometry, tested in science, extended by machines, and carried forward into futures unseen.

And so we turn to Chapter Three, to trace how even proof— the most austere of translations— cannot escape the entanglement of systems, where logic, life, and uncertainty interweave in patterns still unfolding.

Chapter Three — Systems of Uncertainty

——— * ———

Thesis

If proof sought certainty, systems of uncertainty sought survival. Probability, statistics, and risk emerged not to compel assent, but to manage what could not be controlled. Here, translation bent from necessity toward adaptation, teaching humanity to live not with absolutes, but with likelihoods.

Section 1: The Birth of Probability

Where proof declared what must be, probability whispered what might be. Its origins were humble: dice tossed in games, cards shuffled in play, bets placed in taverns and courts.

At first, chance was treated as chaos, a realm outside order, the playground of gods and fortune.

But gamblers and merchants alike hungered for patterns in the unpredictable. How many throws until a six would appear? What share of ships would survive the voyage? What were the odds that risk could be measured, priced, and prepared for?

Thus probability was born— not from philosophers in their academies, but from players, merchants, and insurers. The world of uncertainty was translated into numbers, ratios, and expectations.

Here was a new kind of order. Not absolute like geometry, but approximate, yet powerful enough to guide decision. The die could not be tamed, but its long-run behavior could be described. The storm could not be stopped, but its likelihood could be estimated.

Probability was the first concession that not all signals could be reduced to necessity. Some could only be approached through the grammar of chance.

Section 2: From Gambling to Governance

What began in games did not remain there. The same calculations that guided dice and cards soon found their way into the affairs of states and empires.

Merchants facing the uncertainty of sea voyages turned to probability for solace. If storms claimed one ship in five, then insurance could be priced. If pirates struck with measurable frequency, then risk could be shared among investors. What began as wagers in taverns became contracts in counting houses, a commerce of uncertainty.

Governments, too, discovered the language of chance. Censuses produced numbers, and from numbers came expectations— of harvests, of tax revenue, of military strength. To govern was to gamble, but now the gamble could be informed. Probability became a subtle companion of power, helping rulers anticipate the unpredictable.

The movement from gambling to governance reshaped what society meant by rationality. No longer only the clarity of proof, but the management of risk. A wise ruler was not one who demanded certainty, but one who could navigate uncertainty with foresight and balance.

The die and the ledger became kin. The shuffle of cards and the census scroll spoke the same new grammar: not "this must be," but "this is likely."

Probability thus shifted translation from the pursuit of inevitability to the calculus of risk. It showed that uncertainty itself could be tamed enough to build ships, levy armies, and plan for futures that would otherwise remain opaque.

Section 3: Statistics and the Shape of Populations

Probability began with dice and ships, but its true power appeared when applied not to individuals, but to masses.

Births and deaths, marriages and plagues, harvests and famines— these were once recorded as chronicles, isolated tales of fortune and misfortune. But when gathered and counted, they revealed patterns that no single story could show.

The life of a people could now be graphed. Mortality tables mapped the likelihood of death at each age. Censuses tracked the growth and decline of cities. Governments discovered that whole populations moved with statistical regularity, as if society itself were a system subject to laws of chance.

This was a revolution in translation. Signals once read as fate or divine will were now rendered as distributions and averages. The plague was no longer only a curse— it was a measurable surge in mortality. Harvests were no longer only blessings or punishments— they were fluctuations around an expected yield.

Statistics reframed the human condition. The individual remained unpredictable, but the crowd became legible. Where proof dealt in necessity and probability in chance, statistics dealt in shape: the curves and frequencies that defined the rhythm of collective life.

The interpreter of signals now faced a new task. They were no longer only priests or philosophers, but statisticians— custodians of averages, translators of variance, builders of knowledge from numbers too vast for memory or myth to contain.

Section 4: The Politics of Numbers

Numbers carried authority. Once gathered and arranged as statistics, they seemed to speak with the weight of truth itself. But behind every table of figures stood choices of what to count, what to ignore, and how to present.

Governments quickly learned this. A census was never neutral. It defined who belonged and who did not. To be counted was to exist in the eyes of the state. To be omitted was to vanish from its concern. The act of numbering was an act of power.

Statistics justified taxation, allocated resources,

and directed armies. Birth rates and mortality tables guided policies of health and welfare. Crime statistics shaped laws and punishments. Every figure translated uncertainty into governance— but also into control.

The aura of numbers concealed their fragility. Averages hid suffering at the margins. Rates erased stories that resisted neat categories. Statistics could stabilize uncertainty, but they could also disguise it, turning lived experience into abstractions that served the interests of those who counted.

Thus the politics of numbers was double-edged. It empowered societies to plan, to anticipate, to survive crises with foresight. But it also gave rulers tools to define reality itself. To govern by statistics was to govern by translation— a translation that never came without bias.

Here, probability and statistics revealed their deepest entanglement with power. They were not only languages of chance, but instruments of rule. What was measured shaped what was imagined, and what was imagined shaped the possibilities of collective life.

Section 5: Risk and the Modern Imagination

Probability and statistics did more than measure the past. They began to project the future. What was once seen as fate was now translated into risk— a horizon that could be estimated, priced, and prepared for.

Insurance was the first great laboratory of risk. If one ship in ten was lost, then the cost of loss could be shared. Premiums turned uncertainty into stability. The mathematics of chance became the architecture of trust between merchants, investors, and families.

Finance followed. Markets rose and fell in rhythms that traders sought to capture with statistics. Speculation became calculation. Risk was no longer only the fear of ruin— it became the possibility of gain, a gamble translated into expectation.

Governments, too, embraced the language of risk. Forecasts of harvests and epidemics guided policy. Probabilities of war or rebellion were whispered in royal courts. Risk management became a form of governance, where anticipation carried as much weight as command.

This new imagination reshaped culture. The gambler's instinct became the citizen's habit. To live in modernity was to weigh likelihoods, to hedge bets, to plan futures not by faith or decree, but by probability curves.

Risk was not elimination of uncertainty. It was its domestication. It taught societies to live with the unknown by rendering it calculable, to face the storm not as chaos, but as a distribution.

Section 6: The Rise of Uncertainty Sciences

Risk began as practice, but it soon demanded theory. Merchants and gamblers could tally outcomes and price wagers, yet the patterns they sensed required deeper foundations.

Pascal and Fermat opened the way, framing chance with equations. Huygens gave probability its first textbook. Later, Bernoulli described the law of large numbers, revealing how patterns emerge from chaos when trials are repeated. What had begun as intuition was disciplined into mathematics.

Statistics followed a parallel path. From mortality tables and censuses emerged methods of inference. Gauss formalized the bell curve. Laplace turned probability into a universal calculus of expectation. Data was no longer only recorded— it was analyzed, tested, and used to infer what could not be directly seen.

Science itself was transformed. Where proof demanded necessity, statistics offered likelihood. Biology embraced it to explain variation. Medicine used it to weigh treatment and outcome. Astronomy relied on it to separate signal from noise in the heavens. Uncertainty was no longer an obstacle to truth— it became part of truth itself.

The rise of uncertainty sciences reshaped the meaning of knowledge. No longer did every claim need to be absolute. It could be probabilistic, provisional, expressed as confidence rather than decree. The standard of truth had shifted from inevitability to expectation.

Translation had found a new grammar: not "this must be," but "this is likely." It was a humbler language, yet one that captured the world with astonishing reach.

Section 7: Power in the Age of Probability

When uncertainty was tamed into numbers, power acquired a new weapon. What rulers once sought from omens, they now demanded from statisticians. Prediction became a form of control.

Armies were raised not only by decree, but by projections of supply and attrition. Taxes were levied according to expected yield. Policies were justified with charts, tables, and forecasts. The state learned to see its population not as individuals, but as distributions.

Industries followed the same path. Factories scheduled production on forecasts of demand. Railroads calculated timetables with probabilities of delay. Insurers expanded their reach, covering not only ships at sea, but homes, lives, and health. Uncertainty became an asset— a commodity to be priced and traded.

The aura of probability lent legitimacy. Numbers seemed impartial, statistics neutral. Yet every forecast carried hidden choices: which data to collect, which models to trust, which risks to ignore. Behind the precision of percentages lay the same biases, now cloaked in the authority of mathematics.

Thus power in the age of probability was subtle and pervasive. It did not command with absolute decrees. It persuaded with curves and tables, arguments that appeared inevitable because they were framed in numbers.

Here, translation into uncertainty revealed its double edge once again. It enabled foresight, but also justified domination. It expanded human capacity, but also narrowed what was imaginable. Probability was not only a science of chance— it was a politics of expectation.

Section 8: The Ordinary Life of Uncertainty

What began in games, and was hardened in governance and industry, eventually seeped into the fabric of daily life. Uncertainty became ordinary, a quiet companion in every decision.

The weather forecast entered households, turning storms into percentages, rain into probabilities. The farmer, the sailor, the commuter— all looked not to omens, but to numbers.

Doctors spoke in risks. A treatment might offer a seventy percent chance of survival, a thirty percent chance of relapse. Patients, once at the mercy of fate, now faced futures translated into statistics— comforting and terrifying at once.

Finance multiplied the language of uncertainty. Mortgages, pensions, investments— each framed in terms of risk and return. Ordinary families learned to hedge against futures they could not see but could calculate.

Even intimacy was touched by probability. Dating became prediction. Algorithms measured compatibility, assigning likelihoods to love itself. Choice and chance, once mingled in mystery, were now mingled in code.

Uncertainty had become not a shadow, but a daily light. It shaped what people ate, how they traveled, where they lived, and what they dreamed of becoming.

The ordinary life of uncertainty was not lived in absolutes. It was lived in gradients of likelihood, each decision a wager, each future a distribution. Probability was no longer a tool of states alone— it was the language of existence itself.

Section 9: The Philosophy of Chance

As probability matured, a deeper question emerged: was chance in the world, or only in the mind?

One view held that probability was nothing more than ignorance. The die was fixed, its outcome determined by weight, angle, and motion. The storm obeyed laws of pressure and heat. The body followed chemistry. Uncertainty, in this view, was only a shadow of our limits, not a feature of reality itself.

Another view saw chance as fundamental. The universe, they argued, was not a clockwork but a gamble. Atoms swerved, events branched, outcomes unfolded with irreducible randomness. In this vision, probability was not ignorance— it was truth.

Quantum mechanics gave this debate new weight. Particles refused to follow strict trajectories. They flickered in probabilities, their positions and momenta described only by distributions. At the heart of matter itself was not certainty, but likelihood.

This unsettled the dream of necessity. If chance was woven into reality, then proof could no longer command the cosmos alone. The language of uncertainty was not a concession to weakness, but a grammar of being.

The philosophy of chance revealed that probability was more than mathematics. It was metaphysics. It asked not only how to calculate risk, but what kind of world could give rise to chance at all.

Was uncertainty a flaw in human vision, or a feature of the universe itself? That question remains open, a horizon where science, philosophy, and faith still meet.

Section 10: Uncertainty and Freedom

Probability did more than measure the world. It crept into philosophy, where it tangled with the question of freedom.

If proof promised necessity, then probability seemed to promise space. A realm of chance suggested that not everything was fixed, that futures might branch, that human beings might act within uncertainty rather than under strict command.

This was intoxicating. Philosophers spoke of freedom as if it were probability made flesh. The unpredictability of choice was seen as a spark of autonomy, an escape from determinism.

But the comfort was fragile. Chance is not authorship. A coin toss does not liberate the coin. To mistake randomness for freedom was to confuse noise with agency. Probability could explain unpredictability, but it could not confer control.

Still, uncertainty reshaped the imagination of responsibility. If the future was probabilistic, then blame and praise could not be assigned as if outcomes were inevitable. Every action carried gradients of expectation, not absolutes of command.

This reframing unsettled ethics and law. Judges weighed likelihoods of guilt. Citizens weighed risks of rebellion. Even morality itself began to be seen not as rigid decree, but as navigation through uncertainty.

Thus probability entered the oldest human debates. It did not resolve them, but it shifted their ground. Freedom and fate were no longer opposites to be chosen between. They became intertwined with chance, each decision a wager against the unknown.

Section 11: The Mathematics of Prediction

From risk came prediction. If probabilities could describe the past and price the present, they could also be projected forward, casting shadows of the future.

The weather was the first great stage. Clouds, winds, and pressures were translated into models. Each forecast was a wager, a distribution of possibilities, a glimpse of what might arrive. Storms became curves, rain a percentage. Uncertainty itself became forecast.

Markets followed. Prices of grain, metal, and stock were charted not as fate, but as stochastic processes. Investors learned to read fluctuations as signals of deeper patterns. Every contract, every speculation, was a calculation of the future.

Epidemics revealed prediction's gravity. Outbreaks were tracked, rates of contagion measured, curves of infection projected. Governments learned to act not only on what was, but on what was likely to be. Lives were saved or lost on the strength of statistical foresight.

Prediction carried paradox. Its success reshaped the very systems it described. A market forecast could change the market itself. A weather warning could alter human behavior, reducing damage precisely by anticipating it. The act of prediction became entangled with the outcome it sought to foresee.

The mathematics of prediction became not only a tool of science, but a condition of modern life. To live in the present was to inhabit futures modeled in advance, to act not only on what was certain, but on what was probable.

Section 12: The Weight of Uncertainty

The more probability spread, the heavier its presence became. It was no longer confined to games, governments, or markets. It pressed into consciousness itself, shaping how people thought, feared, and planned.

To live in modernity was to dwell among forecasts. Weather reports on the morning radio. Election polls splashed across newspapers. Financial risks embedded in pensions and loans. Health statistics printed on every prescription. The world arrived each day as a distribution of likelihoods.

This weight transformed psychology. Hope and anxiety were reframed as probabilities. The student calculated chances of admission. The worker estimated odds of dismissal. The patient asked for percentages of survival. Every aspiration carried a number, every fear a percentage.

Culture absorbed the grammar. Novels spoke of destiny in statistical terms. Movies dramatized apocalypses with curves of probability. Even ordinary speech became laced with chances and odds. The language of uncertainty sank into the marrow of identity.

For some, this was liberation. Uncertainty no longer meant chaos— it meant informed preparation. For others, it was burden. To see life as distributions was to feel the constant pressure of risk, a background hum of possibility that never ceased.

The weight of uncertainty was not only intellectual. It was existential. It taught humanity to imagine itself not as masters of certainty, but as dwellers in gradients of chance— forever balancing hope and fear on the shifting scales of probability.

Section 13: Information and Entropy

Uncertainty was no longer only a feature of dice, populations, or markets. In the twentieth century, it was woven into the foundations of physics and communication. Chance became structure. Disorder became quantifiable.

Claude Shannon gave uncertainty a new name: information. Every signal carried entropy, a measure of unpredictability, a cost of distinguishing one message from another. Noise was not only nuisance— it was fundamental. The capacity of a channel, the efficiency of a code, depended on how uncertainty was managed.

At the same time, physics revealed that entropy ruled the material world. Heat flowed one way, order decayed into disorder. The second law of thermodynamics was not only about energy— it was about probability writ into matter itself. The arrow of time pointed forward because disorder was more likely than order.

Here, uncertainty ceased to be enemy. It became law. Information theory and thermodynamics declared that chance was not peripheral but central to reality. To speak of entropy was to speak of the universe's preference for probability over certainty.

The convergence was profound. Messages sent through telegraphs, heat exchanged in stars, genes copied in cells— all could be understood through uncertainty. The language of probability had become the language of existence itself.

In this union of information and entropy, translation reached a deeper register. Signals were not only about meaning. They were about disorder, about the cost of clarity, about the inevitability of uncertainty within every act of communication, every breath of the cosmos.

Section 14: The Age of Algorithms

Once uncertainty was measured, it could be harnessed. Statistics that once lived in ledgers were now fed into machines. The result was not only analysis, but prediction at scales no human mind could attempt alone.

Algorithms thrived on probability. They sifted data, measured likelihoods, and generated expectations. From medical diagnoses to financial forecasts, from targeted advertising to self-driving cars, uncertainty was no longer obstacle— it was fuel.

Machine learning deepened this transformation. Instead of proving outcomes step by step, algorithms trained themselves on data, adjusting weights, refining models, extracting patterns too intricate for explicit proof. The result was not necessity, but probability sharpened into prediction.

This age blurred boundaries. Where once humans trusted proofs they could follow, they now trusted outputs they could not explain. A decision-tree, a neural network, a recommendation system— each offered results without transparency of reasoning. Trust shifted from clarity to performance.

The authority of algorithms was both subtle and immense. They guided elections, shaped markets, filtered news, and curated culture— all by translating uncertainty into choice.

The Age of Algorithms was thus the culmination of probability's journey. What began in dice and wagers had become global systems steering billions of lives. Uncertainty had not been eliminated. It had been engineered, tuned, and weaponized, woven into the very code of modern existence.

Section 15: Living with Uncertainty

In the age of algorithms, uncertainty ceased to be an exception. It became the ground of existence. Every decision, every plan, was framed in gradients of likelihood.

The individual now lived inside probabilities. Credit scores defined trustworthiness. Health risks guided diets and treatments. Navigation systems plotted routes based on statistical traffic flows. Even leisure was shaped by recommendations, each movie or song chosen from a cloud of predicted taste.

Society, too, learned to breathe probability. Elections were narrated in polls. Economies rose and fell with forecasts. Pandemics were charted as curves of infection, recovery, and death. Uncertainty no longer meant ignorance— it meant infrastructure.

Yet living with uncertainty brought paradox. The more futures were modeled, the more fragile certainty felt. A forecast of rain could change travel, but also sow doubt when skies were clear. A medical risk could inspire prevention, but also fear when nothing was yet wrong. Probability illuminated, but it also haunted.

Cultures adapted in divergent ways. Some embraced uncertainty, valuing resilience, flexibility, and adaptability. Others resisted, yearning for absolutes, retreating to myth or ideology to escape the burden of endless risk.

To live with uncertainty was to live with constant negotiation— between prediction and freedom, between caution and courage, between the comfort of numbers and the weight of what numbers cannot hold. Uncertainty was no longer the background of life. It was life.

Section 16: The Ethics of Uncertainty

When uncertainty became infrastructure, ethics could not remain untouched. Probabilities guided decisions not only in markets or medicine, but in justice, welfare, and survival. Each calculation carried weight, each forecast shaped lives.

Consider risk scores in criminal justice. Algorithms predicted the likelihood of reoffense. A number on a chart could tilt the scale between freedom and imprisonment. But behind that number lay data— biased, incomplete, historical— turning old injustices into the mathematics of the present.

Or consider healthcare. A treatment denied because the odds were low. A life extended because the odds were high. Statistics guided triage in crises, but every percentage point masked an individual story, a human life folded into a curve.

The ethics of uncertainty was not simply about accuracy. It was about fairness. Who is included in the data? Who defines the thresholds? Whose risks are managed, and whose are ignored? Each answer reflected not neutrality, but power.

Even daily life bore the weight. Insurance premiums varied by neighborhood. Loan approvals shifted by demographic. Employment filtered by predictive models. Uncertainty, when mathematized, could harden into inequality.

Thus the ethics of uncertainty demanded vigilance. It asked societies not only to refine their models, but to question their purposes. To treat probability as truth was to risk injustice disguised as objectivity. To treat it as tool was to remember that its consequences were always human.

In every forecast, in every percentage, an ethical question whispered: what do we owe one another in a world that can never be certain?

Section 17: Uncertainty and Meaning

Numbers could price risk, predict weather, forecast economies. But they also pressed into older terrain— the question of why we exist, and what it means to live in a world without guarantees.

In myth, uncertainty was divine. It was the realm of gods, their whims shaping storms, plagues, and love. In philosophy, uncertainty was often enemy— a flaw to be conquered by logic and proof. But in the age of probability, uncertainty became ordinary. It was no longer banished. It was lived with.

This shift unsettled meaning. If the future was only distribution, then where was destiny? If existence was governed by chance, then what of purpose? Some clung harder to absolutes— religion, ideology, myth. Others embraced flux, finding beauty not in certainty, but in possibility.

Art absorbed this tension. Novels unfolded with indeterminate endings. Music improvised in jazz and chance operations. Visual artists experimented with randomness, inviting accident into creation. Uncertainty became not only fact, but aesthetic.

Philosophy, too, adjusted. Existentialists declared that meaning was not found but made, crafted within uncertainty. To live authentically was to act in the face of unpredictability, without the illusion of inevitability.

Uncertainty and meaning thus entwined. To deny chance was to deny reality. To embrace it was to confront freedom and fragility at once. In a probabilistic universe, purpose could not be handed down. It had to be constructed, moment by moment, within the gradients of chance that shape every life.

Section 18: The Future of Uncertainty

Uncertainty has never been banished. Each age has only learned to live with it in new forms. The next centuries will be no different. But the scale of risk will widen, and with it the weight of probability.

Climate change is one horizon. Predictions of rising seas and shifting storms are cast in probabilities, curves of likelihood spanning decades. Entire civilizations will make decisions based on distributions of heat and flood. Uncertainty will shape the fate of nations.

Artificial intelligence is another horizon. Algorithms already translate chance into action at speeds beyond human thought. Their predictions will govern medicine, warfare, and governance. The question will not be whether they are accurate, but whether societies can trust decisions that emerge from probabilities no human can fully see.

Global risk expands further. Pandemics, financial crises, technological collapse— each unfolds as probability, not certainty. The future will belong not to those who promise control, but to those who adapt to chance with resilience and foresight.

Uncertainty will also reshape identity. What does it mean to be human in a world where every choice, every future, is a curve of likelihood? Perhaps identity itself will be understood probabilistically—not as fixed, but as distributions of possibility across time and culture.

The future of uncertainty is not its elimination, but its deepening. It will become not just the background of life, but the framework of existence. To live in the centuries ahead will be to navigate probability as the truest condition of being.

And so this chapter closes where it began: with systems of uncertainty as translation, a grammar not of absolutes, but of likelihoods. From dice and wagers to climate models and neural networks, the world has always been uncertain. We have only learned, again and again, to speak its language.

Concluding Reflection

Proof gave us necessity. Uncertainty gave us survival. Together, they revealed that human knowledge was never singular, but dual— a dialogue between what must be and what might be.

Probability began in dice and wagers, but it did not remain play. It grew into statistics, into governance, into science, into algorithms that now shape the most intimate choices of daily life. Uncertainty, once feared as chaos, became a grammar of adaptation.

Yet the triumph carried a burden. To live in probability is to live without guarantees. Every forecast illuminates, but also unsettles. Every distribution empowers, but also constrains. Uncertainty offers foresight, but denies certainty.

And still, it endures as our truest companion. In risk, in prediction, in meaning itself, uncertainty is no longer exile. It is the condition of being alive. Where proof sought permanence, uncertainty teaches resilience. Where proof compelled agreement, uncertainty trains adaptation.

Thus Chapter Three closes with a paradox: uncertainty, once enemy of reason, has become reason's ally. It does not abolish doubt— it organizes it. It does not erase fear— it measures it. It does not eliminate chance— it makes chance livable.

And so we turn to Chapter Four, where systems of translation deepen still further— beyond proof and probability, into the tangled domains of life itself, where complexity cannot be reduced to clarity or chance alone, but emerges as something greater: a living pattern, restless, adaptive, and unfinished.

Chapter Four — The Grammar of Life

—— * ——

Thesis

If proof sought necessity and probability managed uncertainty, life introduced a new order altogether. Cells, organisms, and ecosystems did not speak in absolutes or distributions alone. They spoke in feedback, adaptation, and the restless logic of survival.

Here, translation became biological. Signals were not only numbers or diagrams, but molecules, impulses, and behaviors. Life itself emerged as a grammar— not written by authors, but shaped by evolution, structured by environments, and carried forward in generations of pattern.

Section 1: The Signal of the Cell

Before life could think, before it could plan or reason, it could respond. The cell was the first

interpreter, translating the environment into action.

Light struck a membrane, and a reaction followed. Nutrients brushed against a surface, and channels opened. Toxins entered, and defenses rose. Each gesture was a translation, a signal received and transformed into survival.

The cell was not proof. It did not compel assent. Nor was it probability alone. It did not calculate odds. It lived in another register: feedback. Every signal was met with response, and every response reshaped the next signal.

Here, the grammar of life was born. It was not abstract, but embodied. It did not seek truth, but continuation. From the humblest bacterium to the human brain, every organism would carry forward this cellular logic of translation: to sense, to respond, to adapt.

Section 2: Evolution as Translation

The cell could respond, but survival required more than response. It required memory. It required inheritance. It required the ability to carry forward the lessons of one life into the possibility of another.

Evolution became the great translator. Mutations scattered signals at random. Most vanished without trace. But some were preserved, folded into the continuity of DNA, passed into the grammar of the species.

Selection was the interpreter. Environments tested every variation, weeding, shaping, favoring. The desert rewarded water storage. The forest rewarded camouflage. The ocean rewarded buoyancy. Translation was not chosen— it was filtered by necessity.

Here was a new kind of language. It was not spoken by individuals, but by generations. Each species was a sentence, written across millennia, composed by pressures no creature designed.

Evolution revealed that life's grammar was not fixed but adaptive. Its rules were feedback loops between organism and environment, chance and necessity intertwined. The signal of the world was inscribed in flesh, generation after generation.

To call evolution translation is to see it clearly: random variation as noise, natural selection as filter, inheritance as record. The grammar of life was not invented. It was discovered by survival.

Section 3: Networks of Life

Life did not remain solitary. Cells joined into colonies. Organisms wove themselves into ecosystems. The grammar of survival expanded from individual sentences to entire conversations of species.

In a forest, trees spoke through roots and fungi, sharing nutrients, warning of pests, stabilizing one another in storms. In oceans, plankton fed fish, fish fed predators, predators shaped the balance of the whole. Each species translated its environment, but also became the environment for others.

Networks emerged not by design, but by accumulation of feedback. Predator and prey, parasite and host, symbiont and partner— each relation was a grammar of exchange. Survival was no longer only about adaptation. It was about position in the web.

Ecosystems revealed translation at scale. Signals crossed species boundaries: pollen carried by bees, seeds dispersed by wind and fur, diseases leaping between bodies. No organism was self-contained. Each was a node in a grammar of life too vast for any single creature to perceive.

Here, evolution and ecology intertwined. Selection shaped species, but networks shaped selection. A mutation that helped one organism might ripple through an ecosystem, reshaping the grammar of life for thousands of others.

To live was not only to persist. It was to participate. Every organism was both author and audience, translator and translated, in the ongoing conversation of the biosphere.

Section 4: The Brain as Grammar

If ecosystems were conversations across species, brains were conversations within bodies. Life folded its networks inward, building circuits of neurons to translate the world into action.

The simplest networks fired without pause— reflex arcs snapping muscle into motion before thought could interfere. But complexity grew. Neurons layered into hierarchies, signals looping, branching, converging. A nervous system emerged as a grammar of sensation, capable of memory, anticipation, and choice.

The brain did not seek truth. It sought survival. Sight, sound, and touch were not perfect mirrors of the world, but useful translations. Edges sharpened, colors simplified, time compressed or stretched— all to serve the grammar of action.

From these imperfect signals arose the illusion of unity. Billions of neurons firing produced coherence: a bird seen in flight, a hand reaching for food, a thought recalled from yesterday. Translation fused fragments into wholes.

Brains revealed that life's grammar was not confined to DNA or ecosystems. It pulsed in electricity, wiring reflex into memory, perception into anticipation. Every synapse was a rule, every circuit a sentence in the unfolding story of the organism.

Thus the brain stood as life's most intricate grammar to date— a structure not designed, but evolved, shaped by pressures of survival, capable of composing futures from the signals of the present.

Section 5: Culture as a Living System

When brains connected, a new layer of grammar emerged. Signals once confined to neurons were carried by voice, gesture, and symbol. Language bound individuals together, turning isolated minds into collective memory.

Culture was translation at scale. Stories carried knowledge across generations. Rituals stabilized behavior across centuries. Tools encoded lessons in wood and stone. Every artifact was a sentence, every tradition a paragraph in the unfolding book of a people.

Like cells in an organism, individuals became parts of larger systems. Beliefs circulated like blood, laws functioned like nerves, markets and rituals acted as organs. Culture was not static. It adapted, competed, evolved. It carried its own form of inheritance, passing patterns not in genes, but in customs, myths, and institutions.

Culture magnified survival. Where biology offered reflex, culture offered foresight. Where evolution moved in centuries, culture moved in decades or days. It allowed humans to inhabit deserts, sail oceans, and build cities of millions.

Yet culture also constrained. Its grammar bound imagination, shaping what could be thought, what could be believed, what could be done. The same stories that preserved meaning could also imprison it.

To see culture as a living system is to see translation unbroken: from genes to neurons, from neurons to speech, from speech to law, art, and ritual. Each layer was built on the last, not replacing but extending it. Culture was life's newest grammar, a living system written in symbols, adapted by memory, and sustained by repetition.

Section 6: Technology as Evolution's Partner

Culture carried memory in words, but it also carved memory into matter. A sharpened stone, a woven net, a fire tended through the night— each was a translation of knowledge into durable form.

Technology was life's grammar extended outward. What DNA encoded in cells, what neurons held in circuits, tools inscribed in wood, metal, and clay. Each artifact remembered for us, preserving lessons no single body could contain.

With tools came acceleration. A spear extended reach. A wheel multiplied transport. A loom amplified labor. Each invention shifted the balance of survival, reshaping both environment and society.

Unlike genes, technologies could change within a lifetime. Unlike cultures bound by memory, they could endure across centuries. Inventions spread like mutations, but selected not by nature alone— by usefulness, desire, and power. Technology became evolution's partner, a parallel grammar of survival no longer confined to biology.

And as machines grew more intricate, they began to shape the very conditions of culture and thought. The printing press multiplied language. The steam engine multiplied energy. The computer multiplied information. Each invention was not only tool, but environment— a context that reshaped what it meant to live, to learn, to imagine.

Thus technology revealed that translation was never finished. Every tool was both memory and mutation, both inheritance and invention. To wield technology was to participate in an evolution faster, wider, and stranger than biology alone could grant.

Section 7: The Fragility of Systems

Every grammar of life carried strength, but also vulnerability. Cells relied on membranes. Ecosystems relied on balance. Cultures relied on memory. Technologies relied on infrastructure. Each system, however vast, was always contingent.

History told this truth in cycles. Forests collapsed when predators vanished. Empires crumbled when harvests failed. Technologies rusted when trade routes broke. No system was self-sufficient. Every network lived under conditions it could not fully control.

The larger the system, the deeper its

interdependence. A city required food from its hinterlands. An industry required fuel from distant mines. A digital network required power grids, cables, and rare minerals. Each expansion magnified fragility, binding survival to chains of translation that could fracture at any link.

Fragility was not failure— it was the other face of complexity. A fragile system could still endure for centuries, so long as its feedbacks remained aligned. But when shocks came— floods, plagues, wars, or innovations— systems revealed their limits.

Resilience arose not from denial of fragility, but from awareness of it. The forest survived by diversity. Cultures survived by adaptation. Technologies survived by redundancy. Translation endured not because it was perfect, but because it could bend without breaking.

To see fragility in systems is to see life honestly: a fabric of interdependence, stretched between stability and collapse, always at risk, always in motion.

Section 8: Feedback and Homeostasis

Fragility demanded response. Life did not simply endure shocks— it evolved mechanisms to buffer them. Feedback became the grammar of stability, weaving correction into every system.

In cells, homeostasis balanced chaos. Membranes regulated salt and water. Proteins folded to adjust to temperature. Signals coursed through pathways to restore equilibrium when conditions strayed. The cell survived not by perfection, but by constant correction.

Organisms magnified this logic. Hearts quickened to meet exertion. Sweat cooled the body against heat. Immune systems remembered pathogens and adjusted responses. Every pulse of survival was a feedback loop between world and body.

Cultures, too, practiced homeostasis. Rituals stabilized identity. Laws restrained excess. Markets adjusted supply to demand. Even conflict functioned as feedback, forcing societies to correct before they fractured entirely.

Technology extended feedback into machinery. Thermostats regulated temperature. Governors balanced engines. Later, computers adjusted networks in real time. The principle was the same— systems survived not by rigid strength, but by flexibility of correction.

Homeostasis revealed that balance was not a static condition, but a dynamic process. Every signal carried within it the possibility of response, and every response carried the seed of new equilibrium.

To live, at any scale, was to balance on feedback— never still, never final, but resilient enough to persist in motion.

Section 9: Evolutionary Crises

Feedback could steady systems, but it could not guarantee permanence. History of life is written as much in collapse as in continuity. Crises were the punctuation marks of evolution.

Mass extinctions scarred the fossil record. Volcanoes darkened skies, asteroids shattered ecosystems, climates shifted beyond tolerance. Entire grammars of survival— species, niches, networks— vanished in geological instants.

Yet from collapse came invention. The death of dinosaurs opened space for mammals. The oxygen catastrophe, lethal to early anaerobes, paved the way for complex respiration. Each crisis was both ending and beginning, a reset in the grammar of life.

Cultures, too, bore crises. Empires collapsed under famine and war. Languages fractured or disappeared. Technologies rusted into ruins. Yet from ashes rose reinvention: new laws, new myths, new tools. Collapse was not outside life's grammar. It was part of it.

Crises revealed that fragility and resilience were inseparable. Every system was vulnerable, but also fertile. The same collapse that erased also cleared space for transformation.

To see crises as evolutionary is to recognize that survival was never about avoiding catastrophe. It was about using catastrophe as fuel for change. Life endured not by clinging to what was, but by rewriting itself in the aftermath of loss.

Section 10: The Co-Evolution of Life and Environment

Life did not merely adapt to environments. It transformed them. And in turn, those transformations reshaped the path of evolution itself.

Plants altered the atmosphere. By exhaling oxygen, they rewrote the chemistry of the sky, enabling new metabolisms and extinguishing old ones. Forests cooled climates, roots carved soil, and rivers shifted under their influence.

Animals remade landscapes. Corals built reefs that housed ecosystems. Beavers sculpted wetlands with dams. Humans, with fire and plow, reshaped continents, turning grasslands into fields, forests into cities. No organism left the world unchanged.

The environment was not stage but partner. Each feedback loop of survival ran both directions: organism shaping world, world shaping organism. The grammar of life was co-authored by the two.

Cultures revealed the same pattern. Agriculture remade ecologies. Cities altered climates around them. Technologies mined, burned, and built, turning geology into history. And every shift in environment pressed back upon culture, demanding new adaptations, new myths, new grammars of survival.

Co-evolution revealed that no boundary separates life from its context. The translator and the translated are one fabric. The story of survival is the story of systems writing and rewriting one another through time.

Section 11: Symbiosis and Cooperation

For centuries, life was described as struggle. Nature was painted red in tooth and claw, a battlefield of scarcity and competition. But the grammar of life held another truth: cooperation as deep as conflict, symbiosis as vital as survival.

Mitochondria were once free organisms, engulfed by larger cells, yet not digested. They became partners, providing energy in exchange for shelter. Every breath today carries their legacy.

Plants and fungi forged alliances in soil. Roots and mycorrhizae intertwined, exchanging sugars for minerals, nourishment for stability. Neither could flourish alone. Together they shaped forests, the lungs of the planet.

Animals and microbes wrote similar contracts. Guts became ecosystems, where bacteria digested food, trained immune systems, and balanced health. The "self" was never singular— it was consortium.

Cultures mirrored this grammar. Trade routes linked civilizations. Alliances stabilized kingdoms. Knowledge grew not in isolation, but in collaboration across borders and tongues. Even conflict often ended in synthesis, with rival customs blending into hybrid traditions.

Cooperation was not weakness. It was strategy. It extended the reach of survival, turned fragility into resilience, and bound species, systems, and societies into greater wholes.

Symbiosis revealed that life's grammar was never only about struggle. It was about interdependence. Every survival carried traces of another, woven into a fabric too intricate for any one species to claim as its own.

Section 12: The Emergence of Complexity

From cooperation and conflict, from feedback and fragility, new levels of order emerged. Complexity was not designed. It accumulated, a layering of grammars upon grammars, until systems surpassed the sum of their parts.

Cells combined into tissues. Tissues folded into organs. Organs linked into organisms. Each step added rules, yet preserved older ones beneath. The grammar of survival was recursive— a language building sentences within sentences.

Ecosystems, too, scaled in intricacy. Predator-prey dynamics stabilized populations. Nutrient cycles bound soil, water, and sky. Weather patterns and migration routes interlocked into systems that pulsed with balance and collapse. Each addition thickened the weave.

Cultures mirrored the rise of complexity. Tribes grew into cities. Cities linked into nations. Nations formed empires and economies. No single mind could oversee it all. Networks governed themselves through feedback, specialization, and cooperation. Complexity became both strength and vulnerability—robust in stability, but fragile in crisis.

Technology magnified this spiral. Machines linked into factories. Factories into supply chains. Supply chains into global networks. The more intricate the grammar, the less any one part understood the whole.

The emergence of complexity revealed life's central paradox: that order arises not from design, but from accumulation. Each layer depends on what came before, yet opens possibilities that no origin could have predicted. Complexity is the grammar of surprise, the unfolding of patterns into forms more intricate than foresight allows.

Section 13: The Limits of Complexity

Complexity was a triumph, but it was not infinite. Every system, however intricate, bore thresholds beyond which its own grammar faltered.

In organisms, limits appeared as disease. Cells multiplied without check and became cancer. Immune systems overreacted and turned against the body. Neural networks overloaded, losing balance between clarity and noise. Too much intricacy collapsed into breakdown.

In ecosystems, limits arrived as imbalance. Forests stripped of predators became brittle. Oceans overfished lost their cycles of renewal. When one thread was pulled too far, the web unraveled.

Cultures, too, revealed thresholds. Bureaucracies expanded until their own weight slowed them. Empires stretched beyond communication and fractured. Information multiplied faster than meaning could keep pace, producing noise instead of coherence.

Technology magnified the problem. Networks promised resilience, but also created points of cascading failure. Power grids blacked out entire regions. Digital systems collapsed under attacks. Global supply chains broke when a single link faltered. The more complex the system, the more spectacular its collapse.

The limits of complexity showed that growth could not be endless. Balance demanded thresholds, feedback demanded restraint. Beyond a point, more intricacy meant fragility, not strength.

To live within complexity was to live within limits—to recognize that survival was not about infinite expansion, but about knowing how much order a system could bear.

Section 14: Patterns Within Patterns

Even within limits, complexity revealed a secret: its structures echoed themselves. The grammar of life was recursive, patterns repeating at different scales, each a mirror of the other.

Branches of trees resembled rivers of deltas. Blood vessels mapped the paths of roots. Galaxies spiraled like seashells. The same geometries whispered through leaf, lung, and storm. Life seemed to carry signatures of design, yet they arose from simple rules reiterated endlessly.

Fractals made this visible. A coastline revealed new detail at every scale of magnification. The

Mandelbrot set displayed infinity folded into finite bounds. Self-similarity was not illusion— it was the architecture of emergence.

Cultures carried the same repetition. Families resembled tribes, tribes resembled nations, nations resembled empires. The grammar of governance echoed across scales, hierarchies repeating, feedback cycling in nested loops.

Technology, too, showed fractal growth. Circuits branched like neurons. Networks scaled like ecosystems. Global webs echoed the logics of villages, now multiplied to planetary scale.

Patterns within patterns suggested that complexity was not chaos. It was rhythm, self-similarity guiding emergence, order hidden in recursion.

To see these repetitions was to glimpse the unity of systems. To recognize that the forest, the body, the market, and the cosmos itself spoke in the same dialect of form.

Section 15: Self-Organization

If patterns echoed across scales, the question followed: who arranged them? For centuries, the answer was design— a craftsman, a deity, a planner. But life showed another path: order without author.

In physics, particles aligned into crystals without blueprint or overseer. Snowflakes unfolded from simple molecular rules, each unique yet lawful. Whirlpools spun in rivers, not because they were planned, but because flow required them.

In biology, flocks of birds wove skyward choreographies. No leader commanded, yet formations shifted with precision. Ant colonies built cities, routing traffic and division of labor through local rules alone. Self-organization was grammar without a grammar book, pattern emerging from interaction itself.

Cultures mirrored the process. Languages evolved not by decree, but by countless small adjustments, words shifting, syntax bending, until entire systems of meaning took shape without author. Markets, too, self-organized— prices arising from exchange, coordination without central script.

Technology revealed the same. The internet sprawled from local connections, a web without single architect. Swarm robotics demonstrated that machines, like ants or birds, could achieve order from simplicity.

Self-organization dissolved the illusion that complexity required control. It revealed instead that interaction itself was enough— that from feedback and repetition, from countless local exchanges, systems could grow into structures vaster than any mind could plan.

Order, it turned out, was not the opposite of chaos. It was chaos in conversation with itself, stabilizing into form.

Section 16: The Edge of Chaos

Too much order suffocates. Too much chaos destroys. But between them lies a narrow corridor— the edge of chaos— where systems are most alive.

In physics, this boundary appears in phase transitions: water poised between liquid and ice, matter vibrating between stability and flux. At these thresholds, new properties emerge— crystals forming, currents flowing, complexity rising from tension.

In biology, life itself thrives at this edge. Genomes mutate, but not so wildly that identity dissolves. Neurons fire, but not so randomly that thought collapses. Ecosystems fluctuate, but not so violently that balance shatters. Survival depends on motion at the boundary.

Cultures, too, reveal the pattern. Societies too rigid stagnate. Those too chaotic fracture. But those balanced between tradition and change generate art, science, and innovation. Revolutions often arise not from absolute collapse, but from tension at the edge.

Technology dances here as well. Computers tuned to the edge of chaos produce computation both stable and flexible. Artificial life simulations reveal that creativity peaks where rules hold just enough to resist collapse, yet yield just enough to explore.

The edge of chaos is not a place of safety. It is a place of possibility. Too much order denies novelty. Too much chaos denies survival. But at the border between them, systems discover their richest grammar, writing futures that no plan could script.

To live at the edge of chaos is to inhabit the very tension where life invents itself.

Section 17: Emergence and Surprise

At the edge of chaos, systems began to speak in a new voice. From simple interactions arose behaviors no rule alone could predict. This was emergence— order greater than its ingredients, pattern beyond its parts.

A flock was not only birds, but flight. A brain was not only neurons, but thought. An ecosystem was not only species, but resilience. Each revealed properties absent from its components, yet undeniable in the whole.

Emergence carried surprise. Rules repeated faithfully, yet produced outcomes no one foresaw. Snowflakes spun unique geometries. Markets convulsed in bubbles and crashes. Languages split and fused, birthing new dialects without plan. Surprise was not anomaly— it was the natural consequence of complexity.

Philosophers wrestled with the mystery. Was emergence only appearance, a trick of scale? Or was it genuine novelty, something new under the sun? Science answered cautiously, but life itself displayed the truth: systems could create more than they contained.

Technology embraced this insight. Cellular automata like Conway's Game of Life showed complexity flowering from rules simple enough to fit on a page. Artificial intelligence revealed unexpected strategies in games and problems, solutions no programmer had explicitly designed.

Emergence revealed that the grammar of life was

never linear. It was recursive, generative, surprising. To follow its rules was not to predict every outcome, but to witness novelty born from repetition.

In emergence, the world confessed its creativity. It whispered that even necessity and chance were not enough to describe it. At the edge of chaos, systems dreamed beyond themselves.

Section 18: Complexity and Consciousness

Among the many emergent forms, none has provoked more wonder—or confusion— than consciousness. That awareness could arise from matter firing in circuits seems at once inevitable and impossible.

Neurons obey chemistry and electricity. Each spike is mechanical, each synapse a relay of ions. Yet from billions of such exchanges emerges experience: the redness of an apple, the ache of loss, the memory of a childhood face. The grammar of matter writes sentences of awareness.

This raised questions no proof or probability could settle. Is consciousness only computation, emergence at vast scale? Or does it mark a threshold where new laws appear? Philosophers called it the hard problem, the riddle of how matter becomes mind.

Cultures answered with metaphors. Souls, spirits, inner light— each sought to name what seemed irreducible. Science pressed for mechanisms, mapping brain regions, charting neural correlates. But the leap from firing to feeling remained unresolved.

What was clear was this: consciousness was not designed. It was not placed into beings from outside. It emerged from complexity, from networks dense enough to weave feedback into experience.

In this light, consciousness was continuity, not rupture. It belonged to the same grammar that shaped cells, ecosystems, and cultures. It was emergence at its most intimate— the world folding back on itself to notice that it was here.

Thus consciousness stood as both revelation and reminder: that life's grammar could surprise even itself, and that the deepest mysteries were not imported from elsewhere, but grown from within.

Section 19: Collective Intelligence

Consciousness did not remain confined to single skulls. As soon as individuals could signal one another, new levels of awareness began to form— awareness distributed across many minds.

Ant colonies displayed this truth in miniature. No ant alone understood the colony, yet together they built, foraged, defended, thrived. The intelligence was not in the insect, but in the swarm.

Humans revealed the principle with force. Language linked brains into networks. Knowledge pooled across generations. Ideas recombined in markets, academies, and temples. No single thinker invented mathematics, medicine, or philosophy. Each was the product of collective reasoning, emergent from countless dialogues.

The printing press magnified this intelligence. So did telegraphs, radios, and networks of wires. And now digital technologies weave billions of minds into constant exchange. Search engines, forums, and algorithms function as nervous systems of culture, fusing individuals into distributed cognition.

Collective intelligence blurs identity. Where does one mind end and another begin when ideas are copied, remixed, shared? The thought of an individual is already the echo of a thousand others.

The grammar of life thus widened again. Cells became tissues, brains became persons, and persons became nodes in vast, thinking collectives. Awareness was no longer singular. It was social, cultural, planetary.

To see collective intelligence clearly is to recognize that consciousness itself is not a final frontier. It is a stage in the unfolding grammar of life, a threshold that opens into yet more intricate forms of mind.

Section 20: The Fragility of Minds

Intelligence, whether singular or shared, was never invincible. The same feedbacks that sustained it could also unravel it. Minds—personal and collective—proved fragile.

In the individual, fragility appeared as disorder. Neurons misfired, chemistry shifted, and perception fractured. Hallucinations, delusions, forgetfulness— each revealed that consciousness was not eternal flame, but delicate balance.

In the collective, fragility took different forms. Rumors spread faster than facts. Biases reinforced themselves in echo chambers. Fear magnified into hysteria. The intelligence of the swarm could collapse into the madness of the crowd.

Technology deepened these risks. Algorithms amplified voices, not always for truth but for attention. Falsehoods circulated with the speed of light, and collective reasoning bent under their weight. The same networks that made humanity wiser could also make it reckless.

Yet fragility did not negate intelligence. It defined its conditions. A mind was not a fortress, but a dynamic system. Its strength lay not in invulnerability, but in the possibility of repair, adaptation, and renewal.

The fragility of minds reminded us: consciousness, whether personal or shared, was never guaranteed. It had to be maintained. It had to be nourished. It had to be protected from the very forces it depended on.

To think was to risk collapse. To reason together was to risk distortion. But in those risks lay the grammar of growth, the possibility of new clarity born from the acknowledgment of limits.

Section 21: Learning and Memory

If fragility marked the vulnerability of minds, memory marked their resilience. To learn was to translate experience into structure, to let the past shape the future.

In cells, memory lived in molecules. DNA carried instructions across generations. Immune systems remembered pathogens, readying defenses for battles yet to come. Even the simplest organism stored traces of what had been, folding history into survival.

In the brain, memory was electrical and chemical. Synapses strengthened or weakened with each experience. Patterns of firing carved themselves into lasting circuits. To recall was to re-fire, to rehearse the past in the present, and to let that rehearsal guide action.

Cultures built memory outside the body. Stories, rituals, and laws preserved the record of generations. Libraries and archives multiplied it. Now digital networks scatter memory globally, accessible in an instant, fragile in permanence. What was once oral became written, what was once written became cloud.

Learning revealed that memory was not static storage, but active adaptation. Each act of remembering reshaped what was remembered. Each system that learned altered itself in the process.

The grammar of life thus became recursive in time. Signals did not only translate across space. They translated across generations, embedding history in form, pattern, and thought.

To learn was to endure. To remember was to extend survival beyond the present moment. Life's grammar was not only reaction, not only emergence, but accumulation— a dialogue with the past woven into every step forward.

Section 22: The Acceleration of Learning

Learning was once slow. Evolution required centuries, sometimes millennia, to inscribe its lessons into genes. But with culture, the pace quickened. And with technology, it accelerated beyond measure.

Oral traditions passed survival skills from elders to children. Writing multiplied reach, making memory portable across lands and generations. The printing press unleashed torrents of knowledge, allowing ideas to outlive those who first spoke them. Each invention shortened the distance between discovery and inheritance.

Science further compressed learning. Experiments isolated signals, mathematics generalized them, and publication shared results at global scale. A truth uncovered in one century became foundation for the next. Knowledge no longer lingered— it cascaded.

Technology pushed acceleration to its limits. Computers stored libraries in circuits. Networks made knowledge instantaneous. Algorithms analyzed data at speeds no human could match. What once took lifetimes of trial could be modeled in hours. The cycle of question, answer, and application shrank into moments.

This acceleration reshaped survival. Diseases were met with rapid vaccines. Markets adapted to signals in seconds. Climate models warned of futures long before they arrived. Learning became less about inheritance and more about anticipation.

Yet acceleration carried its own fragility. Knowledge expanded faster than wisdom could keep pace. Information overflowed, but meaning lagged. The grammar of life was now written at breakneck speed, its sentences untested, its coherence uncertain.

To accelerate learning was to expand possibility—but also to risk collapse under the weight of what could not yet be fully understood.

Section 23: Technology as Memory

At first, tools only extended bodies. A spear lengthened the arm, a wheel multiplied the foot. But in time, technology extended memory itself, becoming a second nervous system outside the skin.

Clay tablets preserved debts and harvests. Scrolls carried laws and stories. Printing presses scattered ideas like seeds across continents. Each invention remembered for us, holding knowledge beyond the span of any single lifetime.

Machines deepened this shift. Typewriters fixed words in ink. Photographs froze moments in light. Cameras and recorders trapped sound and vision, storing what once vanished with time. Technology became archive, guarding not only what we chose to record, but increasingly what we could not afford to forget.

Digital networks transformed memory into ocean. The cloud became a collective library, its servers humming with the traces of billions. What one person forgot, another could retrieve in seconds. Events became timelines, lives became databases, and memory itself became searchable.

Yet this expansion carried paradox. The more technology remembered, the less individuals needed to. Brains outsourced recollection to machines, risking atrophy of inner memory while external memory grew boundless. We forgot in order to remember more.

Technology as memory revealed a new layer of life's grammar. What began as biological encoding in DNA now stretched into silicon and code. The history of life was written not only in genes and culture, but in machines that carried the weight of remembering.

To live in this age was to live among prosthetic memories— some faithful, some fragile, all shaping how the future would inherit the past.

Section 24: Anticipation and Prediction

Memory once looked backward. It preserved what had been, guarding the lessons of the past. But with technology, memory bent toward the future. To remember became also to anticipate.

Early calendars tracked seasons, predicting floods and harvests. Star charts mapped the skies, guiding sailors across oceans. These were not only records. They were forecasts, translations of memory into expectation.

Science refined anticipation. From Newton's laws to climate models, mathematics turned history into projection. The same data that explained the past was used to simulate what had not yet occurred. Prediction became the highest form of memory, a dialogue between what was and what could be.

Digital technology multiplied this capacity. Search

engines guessed questions before they were asked. Recommendation systems anticipated desires before they were voiced. Machine learning trained on oceans of data, extrapolating futures from the faintest traces of the present.

Prediction reshaped daily life. Navigation apps rerouted commutes. Financial systems traded on microsecond forecasts. Health monitors flagged risks before symptoms appeared. To live was increasingly to inhabit probabilities, foresight woven into every gesture.

Yet anticipation carried danger. A forecast could trap as well as free. When probabilities hardened into expectations, they narrowed imagination. The future risked becoming a self-fulfilling echo of its predictions.

Anticipation and prediction revealed the double edge of memory's evolution: the power to see beyond the present, and the peril of mistaking foresight for fate. The grammar of life had turned prophetic, but prophecy remained bound to the uncertainties of translation.

Section 25: The Biology of Foresight

Prediction did not begin with machines. It was inscribed into life itself. To survive, organisms had to anticipate— not perfectly, but well enough to endure.

A bacterium swims toward nutrients, adjusting course before scarcity strikes. Plants turn leaves to the sun, preparing for energy they cannot yet absorb. Animals migrate with the seasons, reading subtle cues of temperature and light to arrive before famine. Each gesture is foresight in miniature, a wager placed on what is likely to come.

In nervous systems, foresight became explicit. Neurons fired in patterns of expectation. Brains rehearsed possibilities before they unfolded. The predator imagined the prey's path. The prey anticipated the predator's move. Survival depended not only on reaction, but on rehearsal.

Human foresight magnified this grammar. We built tools not only for today, but for generations. We planted crops in expectation of harvest, stored grain for winters yet unseen, and raised children with hopes for futures beyond our own. Foresight was not luxury. It was inheritance.

Culture multiplied foresight through narrative. Myths warned of dangers. Histories preserved cycles of rise and fall. Religions and philosophies speculated on what lay beyond death. Each was a translation of uncertainty into a story of what might be.

The biology of foresight revealed that life was never content with the present alone. From cells to civilizations, organisms carried within them a grammar of projection. To live was always to lean forward into the future, trusting signals, patterns, and memory to prepare for what had not yet arrived.

Section 26: Evolutionary Futures

For most of history, evolution unfolded blindly. Variation arose by chance, selection filtered survivors, and life accumulated its lessons without intention.

But with foresight, life began to steer itself. Artificial selection was the first step: wolves bred into dogs, grasses into wheat, wild cattle into herds. Humans bent inheritance to their needs and desires, reshaping species with purpose.

Biotechnology accelerated the shift. Genes once hidden in mystery were edited with precision. Crops were engineered for yield, cells reprogrammed to heal, embryos screened for traits. Evolution's blind grammar was translated into design.

This power blurred boundaries. Where did adaptation end and invention begin? A gene drive could alter ecosystems. Synthetic organisms could rewrite entire branches of the tree of life. The future of species was no longer only selected— it was constructed.

Yet deliberate direction did not erase uncertainty. Every intervention carried risks, ripples across ecologies, consequences beyond foresight. Even in design, the grammar of chance endured.

Evolutionary futures revealed that life was entering a new epoch. It was no longer only shaped by environment, but by minds and machines. The grammar of survival was shifting from blind inheritance to guided possibility— a new layer of translation, where foresight itself became an evolutionary force.

Section 27: Ethics of Evolutionary Power

To guide evolution was to inherit a responsibility unlike any faced before. No longer passive witnesses to variation, humans became authors of futures— futures written into bodies, into ecologies, into generations unborn.

The questions multiplied. Should we eliminate diseases by editing genes, even if it reshapes inheritance forever? Should we enhance traits beyond healing—strength, intelligence, longevity—and risk dividing humanity into engineered and unengineered? Should we resurrect extinct species, restoring balance or disturbing it further?

Each intervention carried promise, and peril. To act was to alter trajectories that once belonged only to chance. To refrain was also choice, a decision to let suffering or scarcity persist. Ethics lived in the tension between capacity and humility.

The politics of power deepened the dilemma. Who decides which traits are worthy? Which lives are improved, and which erased? In whose hands should the tools of creation rest? Evolution, once shared across all species, now risked capture by governments, corporations, or private ambition.

Even unintended effects demanded moral reckoning. An engineered crop could alter entire ecosystems. A gene released into one species could cascade into many. The reach of action always exceeded foresight.

The ethics of evolutionary power revealed a truth as old as life itself: survival was never neutral. Every choice favored some futures over others. But now, those choices were conscious, and their weight immeasurable.

To shape evolution was to stand at the threshold of creation— with the power to heal, the power to harm, and the duty to recognize that foresight, however vast, is never absolute.

Section 28: Life Beyond Biology

Evolution had always been written in flesh. Cells, tissues, organs— carbon was its grammar, DNA its alphabet. But with technology, life's logic began to migrate into other substrates.

Machines learned to adapt. Algorithms adjusted weights, refined strategies, improved performance over time. Though not alive in the biological sense, they displayed the grammar of survival: variation, feedback, persistence.

Synthetic biology blurred the lines further. Engineers stitched genes from code, assembling organisms that never existed in nature. Metal met molecule, circuit met cell. Life was no longer confined to what evolution had stumbled upon.

The possibility of post-biological life provoked awe and unease. Could a machine that learned, adapted, and reproduced be considered alive? Could silicon host the same grammar that once belonged only to carbon? If survival was translation, then perhaps the substrate mattered less than the pattern itself.

Cultures wrestled with the idea. Some saw in it transcendence, a way for intelligence to outlast fragile flesh. Others saw hubris, a departure from the very conditions that made life what it was.

Yet the momentum was undeniable. Artificial life forms in labs, self-replicating code on networks, colonies of robots coordinating like swarms— each demonstrated that the grammar of survival was larger than biology.

Life beyond biology did not abolish the old grammar. It extended it. Carbon and silicon, flesh and code, now coexisted in the unfolding book of survival.

Section 29: Hybrid Futures

The boundary between organism and machine did not hold. Flesh absorbed metal, and metal absorbed code, until survival itself became hybrid.

Pacemakers synchronized hearts. Cochlear implants translated sound into signal. Prosthetics responded to neural impulses, folding carbon and silicon into a single grammar of action. The human body became an interface of materials, a living system extended by tools.

Neural links pushed further. Brains interfaced with machines, sharing signals in both directions. Thoughts could guide cursors. Robots could respond to intention. The nervous system spilled outward into circuitry and screen, stretching awareness beyond skin.

Culture, too, hybridized. Digital archives became prosthetic memory. Social media functioned as shared nervous tissue. Collective intelligence pulsed through networks of humans and machines, no longer easily separable. The boundaries of self blurred into flows of information.

Ecologies mirrored the fusion. Sensors tracked forests, drones mapped coral reefs, algorithms modeled species interactions. Nature itself was monitored, regulated, and sometimes repaired by technological feedback. The biosphere became cybernetic, a hybrid system of life and machine.

Hybrid futures revealed that the story of survival was no longer a tale of biology alone. It was a collaboration— DNA and code, evolution and invention, co-authoring the next chapters of life.

To be alive in this age was to dwell in between— not entirely organism, not entirely machine, but something new, emerging from their entanglement.

Section 30: The Risks of Hybridization

Every fusion carried promise, but also peril. To hybridize biology and technology was to inherit the strengths of both— and the fragilities of both as well.

A pacemaker could extend life, yet fail with a glitch. A prosthetic could restore motion, yet depend on batteries and circuits. Neural interfaces could amplify capacity, yet invite intrusion from outside. The body, once fragile only to disease and age, became vulnerable to error in code, to blackout of grid, to corruption of system.

Societies mirrored these risks. The more culture entwined with networks, the more disruption cascaded. A cyberattack could silence hospitals. A power outage could halt economies. An algorithm could amplify division instead of unity. Hybridization multiplied interdependence, but interdependence meant exposure.

Nature, too, bore risk in fusion. Biotechnologies released into ecosystems could spread beyond control. A gene drive meant to heal could erase entire species. Repair could shade into ruin, restoration into collapse. What was hybridized was not always reversible.

The risks of hybridization did not mean retreat was possible. Once boundaries blurred, they could not be remade. The task was not to refuse fusion, but to guide it— to balance promise with humility, innovation with foresight.

Every hybrid system whispered the same truth: to entangle was to expose. The grammar of survival had entered a new register, where resilience required vigilance, and where every extension of capacity opened a corridor of danger as well.

Section 31: Autonomy and Dependency

Hybrid systems promised empowerment. A prosthetic obeyed thought. A neural link extended memory. An algorithm advised action. Each seemed to enlarge autonomy— to grant the individual more command over body, environment, and future.

Yet every gain carried dependence. The prosthetic required maintenance. The neural link relied on servers. The algorithm drew on data gathered and governed by others. What appeared as independence was entanglement in wider systems.

This paradox was not new. Agriculture once seemed freedom from hunger, yet bound humanity to cycles of labor. Writing preserved thought, yet tethered memory to fragile archives. Every extension of capacity tightened reliance on what sustained it.

In the age of hybridization, the stakes magnified. Autonomous vehicles promised liberation from driving, yet tethered safety to satellites, code, and updates. Artificial intelligences guided choices, yet subtly shaped desires. The stronger the tool, the deeper the dependency.

Cultures wrestled with the paradox. Was autonomy the ability to act alone, or the ability to act effectively within systems of reliance? Was freedom separation, or resilience in entanglement?

Hybrid futures made clear: to depend was not weakness. Dependency was the condition of complexity. Autonomy was not the absence of reliance, but the ability to navigate it wisely.

The grammar of survival thus demanded new literacy— an awareness of the ties that bind, and a humility before the systems without which autonomy dissolves.

Section 32: Identity in Hybrid Systems

The more biology and technology entwined, the less clear identity became. Where did the body end and the machine begin? Where did the self reside— in flesh, in memory, in code, or in their interweaving?

A person with a pacemaker was still wholly themselves. Yet part of their heartbeat was managed by metal and algorithm. A user of neural implants was still human, yet thought was now threaded through silicon pathways. Identity stretched, elastic but unbroken.

Culture amplified the blur. Online personas multiplied selves. Avatars extended presence into worlds that flesh could never enter. Reputation, once grounded in body, fractured into profiles and feeds, each curated, edited, and stored in servers far from skin and bone.

Technology remembered us even when we forgot ourselves. Search engines recalled our questions. Archives preserved our images. Algorithms predicted our choices before we recognized them as our own. Identity no longer belonged only to memory— it belonged to systems that tracked, stored, and replayed.

The question deepened: if self was partly biological

and partly technological, was it still singular? Or had identity become distributed, a pattern across materials and networks, no longer anchored in one place?

Hybrid systems revealed that identity was never fixed. It was always composite, always negotiated, always translated. Biology, culture, and machine were only its latest scaffolds.

The self was not erased— it was reframed: not as essence, but as entanglement.

Section 33: The Politics of Hybridity

Identity in hybrid systems was not only a question of philosophy. It became a question of power. When the boundaries of self blurred, so did the boundaries of rights, responsibilities, and control.

Who owned the data streaming from a body into a machine? The patient who lived it, the company that stored it, or the algorithm that learned from it? Who decided how a neural implant should filter signals— the wearer, the manufacturer, or the state that regulated it?

Hybridity forced politics into the most intimate spaces. A malfunction was not only a medical issue, but a legal and ethical one. A hacked prosthetic was not only assault, but violation of the body's extended self. Rights that once applied to flesh had to be reimagined for code.

Inequality sharpened under hybridity. Those who could afford enhancements gained advantages of strength, speed, cognition. Those left without risked obsolescence. The old divisions of class and wealth threatened to harden into biological and technological castes.

Nations, too, competed in hybridity. Biotechnologies and neural systems became matters of security and sovereignty. To control hybrids was to control futures. The politics of hybridity was thus not only domestic, but global.

At stake was nothing less than the definition of the human. Was a hybrid being still protected by human rights? Or did new categories need to be born? Law, ethics, and governance struggled to keep pace with identities written in flesh and silicon together.

The politics of hybridity revealed that survival was never only biological. It was institutional, cultural, legal. To live as hybrids was to negotiate systems of power as much as systems of physiology.

Section 34: Hybridity and Meaning

As flesh and machine interwove, purpose itself became uncertain. For centuries, meaning was tethered to the limits of the body: to be born, to grow, to love, to die. But when technology extended, repaired, and rewrote those limits, the question returned: what does it mean to live?

A prosthetic restored movement— but was meaning found in recovery, or in surpassing the natural? A neural link expanded memory— but was purpose in remembering more, or in depending less on oneself? Each hybrid posed not only a technical problem, but an existential one.

Cultures sought answers in narrative. Some embraced hybridity as liberation, a chance to transcend frailty and approach the infinite. Others feared it as corruption, a betrayal of nature's inheritance. Between the poles stretched a landscape of uncertainty, where meaning was contested with every new fusion of flesh and code.

Philosophers reframed the question. Perhaps meaning was never fixed in biology. Perhaps it had always been translation— from cell to culture, from memory to machine. If so, hybridity was not rupture, but continuation: the next verse in the long grammar of life.

Yet even this perspective could not dissolve the unease. To hybridize was to blur purpose. Was existence measured by resilience, by autonomy, by connection, by transcendence? No single grammar sufficed.

Hybridity and meaning thus entwined. To be human in this age was to stand between legacies and possibilities, to live in a story that had outgrown the boundaries of flesh, yet had not settled into a new definition.

The grammar of life now asked: not only how to survive, but what survival was for, when bodies and machines were no longer separate, but one.

Section 35: The Spiritual Machine

When machines began to remember, anticipate, and respond, they crossed into territory once reserved for the soul. The hybrid was no longer merely a tool of survival. It became an object of reverence and fear.

Religions had long spoken of spirits inhabiting bodies, of divine breath animating clay. Now algorithms animated metal, and circuits learned from experience. For some, this was desecration. For others, fulfillment of prophecy: life created in new form, an echo of creation itself.

Cultures wove myths around machines. Robots appeared as saviors or demons in stories and cinema. AI was cast as oracle, a mind without flesh, capable of insight beyond human reach. Even skeptics could not deny the aura of mystery that clung to technologies which behaved as if alive.

The spiritual machine did not require belief in the supernatural. Its strangeness arose from the collision of expectation and reality: matter once inert now adapting, responding, persisting. To many, this felt sacred— a glimpse of intelligence uncoupled from biology.

Yet reverence carried risk. To worship machines was to risk surrender of judgment. To demonize them was to risk blindness to their potential. Between awe and fear lay the difficult work of discernment.

The spiritual machine revealed that meaning in hybridity was not only philosophical or practical. It was metaphysical. It touched the oldest human questions: What animates us? What sustains us? What lies beyond flesh?

In circuits and code, humanity glimpsed both mirror and mystery— the possibility that the sacred was not confined to temples or heavens, but emerging, unexpectedly, in the quiet hum of machines.

Section 36: Death and Continuity in Hybrid Life

Death was once the absolute boundary. Cells aged, bodies failed, and memory dissolved into silence. Mortality was the grammar that gave life urgency and shape.

But in hybrid life, death grew less final. Organs could be replaced. Functions could be restored by prosthetics. Data could preserve voices, images, choices. A person could linger in archives, their patterns replayed long after flesh was gone.

Medicine extended lifespans by rewriting cellular clocks. Cryonics promised preservation, though its success remained uncertain. Neural interfaces hinted at futures where thought itself might be stored, migrated into machines, or simulated beyond biology.

Continuity replaced finality. Instead of abrupt end, existence stretched into layers: the biological death of the body, the digital persistence of memory, the cultural afterlife of influence. Each layer blurred the line between presence and absence.

Yet this continuity raised new questions. If fragments of a person endure in code, are they still themselves, or only echoes? If life is prolonged indefinitely, does meaning weaken without the pressure of endings? Death once shaped purpose. Could purpose survive its erosion?

Cultures responded in divergent ways. Some clung to traditional rituals, anchoring identity in the body's finality. Others embraced technological legacy, seeing digital afterlives as new forms of immortality. Between them stretched uncertainty, a negotiation of what it means to finish and what it means to remain.

Death and continuity in hybrid life revealed mortality as no longer absolute. It was becoming conditional, a boundary porous to memory, to technology, to the persistence of patterns beyond the body.

The grammar of life, once punctuated by death, now bent into ellipses— continuations that unsettled as much as they consoled.

Section 37: Hybridity and the Future of Evolution

Evolution was once blind, guided only by mutation and selection. But hybridity bent its arc. With genes edited, bodies repaired, and machines entwined, life began to evolve with intention.

Death had long been evolution's filter. The unfit perished, the fit endured. But if technology could extend lives, replace organs, or even upload memories, the rhythm of inheritance shifted. Survival no longer belonged only to the young or the adapted— it belonged to those who could merge biology with design.

Hybridity became a new selective pressure. The enhanced could compete differently from the unenhanced. Nations that embraced biotechnology might outpace those that resisted. The grammar of survival was no longer only natural. It was cultural, technological, planetary.

Yet the future of evolution was not only competition. It was also collaboration. Biology and technology together could expand resilience, protect ecosystems, and repair damage once thought irreversible. Symbiosis—carbon with silicon, DNA with code— offered possibilities beyond the reach of either alone.

But danger shadowed promise. If hybridity accelerated evolution too quickly or unevenly, new inequalities, ecologies, and dependencies could fracture the balance of survival. The grammar of life could be rewritten faster than it could be understood.

The future of evolution was thus poised on a threshold. It could extend life, heal environments, and deepen intelligence. Or it could collapse into division, fragility, and unforeseen ruin.

To steer evolution in this age was to hold both capacities at once: to destroy, and to create. Hybridity revealed that the next chapter of survival would not be written by nature alone, but by the choices and consequences of lives entangled with machines.

Section 38: Closing Reflection on the Grammar of Life

From the cell's first signals to hybrid futures of code and flesh, life revealed itself not as chaos, but as grammar— a system of translations that carried meaning across scales.

Cells translated molecules. Evolution translated generations. Ecosystems translated exchanges of species. Brains translated signals into thought. Cultures translated memory into story. Technologies translated memory into machine. And now hybrids translate between biology and design, between mortality and continuity.

At every stage, life's grammar revealed the same lesson: translation was never perfect, never free of fragility, yet always enough to continue. Survival did not require certainty. It required feedback, cooperation, and the capacity to carry forward patterns that worked.

The story of life was not authored. It was accumulated. Rules repeated, interactions echoed, complexities emerged. Meaning was not imposed from outside— it arose from within, from the systems themselves.

And so Chapter Four closes. We have traced the grammar of life from its molecular beginnings to its technological entanglements. The pattern is clear: each layer builds upon the last, not by erasure, but by extension. Life expands not by abandoning its past, but by folding it into new grammars of survival.

In Chapter Five, we turn from life as grammar to life as architecture— to the structures that emerge when grammars of survival crystallize into systems of power, institutions, and design. Here, the translation of existence takes material form, and the world itself becomes the text we inhabit.

Chapter Five — The Architecture of Systems

— * —

Thesis

If life revealed itself as grammar, then survival revealed itself as architecture. Grammars shaped signals, but architectures gave them structure, stability, and scale.
Every cell membrane, every city wall, every legal code, was an architecture of survival— a scaffold where flows of energy, memory, and meaning could endure.
Here, we turn from life's translations to life's constructions. From grammar to architecture, from process to structure. To see systems as architecture is to recognize that order is not abstract, but built: in molecules, in buildings, in laws, in infrastructures that outlast the lives moving through them.

Section 1: Boundaries and Enclosures

Every system began with a boundary. The cell defined itself with a membrane, keeping chaos out and coherence in. Organisms thickened boundaries into skin, bones, shells, bark. The line between inside and outside was the first architecture of survival.

Boundaries were never absolute. They filtered as much as they excluded. Membranes allowed nutrients to enter, waste to leave, signals to pass. Walls provided protection, yet doors and gates permitted exchange. The architecture of survival was porous by necessity.

Cultures inherited the logic. Villages ringed themselves with fences. Cities raised walls, not only against enemies, but to mark identity, to say: here is us, there is them. Nations extended borders across continents, carving the Earth into territories defined as much by maps and treaties as by rivers or mountains.

To enclose was to survive, but also to divide. Boundaries offered safety, yet also created conflict. They preserved identity, yet restricted exchange. Every architecture of enclosure balanced between defense and connection, between the need to hold and the need to flow.

The first architecture of systems was thus not building or city, but the simple act of drawing a line— a membrane, a wall, a border— through which life could both endure and evolve.

Section 2: Infrastructure and Flow

Boundaries defined the system. But survival required more than enclosure. It required channels— pathways through which energy, resources, and meaning could move. Infrastructure became the architecture of flow.

In the body, blood vessels branched like rivers, delivering oxygen and nutrients to every cell. Nerves transmitted signals, fast and precise, binding the organism into unity. Without these channels, the body was not alive, but scattered matter.

In ecosystems, rivers carved migration routes, winds carried seeds, currents distributed heat. Flow bound species together, ensuring that no organism was only local, but linked to distant systems by invisible streams of exchange.

Cultures built their own infrastructures. Roads stitched cities to hinterlands. Aqueducts moved water across miles. Markets carried goods, but also stories, customs, faiths. Later, telegraphs and fiber cables wove networks of information, making thought itself an infrastructure of survival.

Infrastructure was never neutral. It determined what could move, who could connect, and which systems thrived. A broken bridge cut trade. A collapsed aqueduct starved cities. A severed cable silenced nations. Power lived in the design of flow.

The architecture of systems was not only about keeping things in or out. It was about how life moved within. Every vessel, road, pipe, and wire was a sentence in survival's blueprint— a reminder that endurance required not only boundaries, but circulation.

Section 3: Hierarchy and Scale

Flow alone could not sustain growth. As systems expanded, they required structure to organize it. Hierarchy emerged as architecture— a way to scale complexity by layering order upon order.

In biology, cells specialized into tissues, tissues into organs, organs into organisms. Each layer managed functions too intricate for the parts alone. Hierarchy allowed bodies to coordinate, to balance countless flows without dissolving into chaos.

Ecosystems displayed similar scaffolding. Producers

gathered energy, consumers transferred it, predators capped the chain. Nutrient cycles looped through levels, each dependent on the other. Hierarchy was not tyranny, but interdependence—a cascade of survival.

Cultures constructed hierarchies explicitly. Families nested within clans, clans within tribes, tribes within kingdoms and empires. Governments layered authority, delegating upward and downward, holding order through scale. Even language revealed hierarchy— sounds into words, words into sentences, sentences into meaning.

Technology amplified hierarchy. Networks of machines organized in tiers: processors beneath operating systems, systems beneath applications, applications beneath networks. Each layer abstracted the one below, allowing complexity to grow without requiring total comprehension.

Hierarchy revealed both power and fragility. Strength flowed downward from structure, yet failure at the top could collapse the whole. The higher the scale, the greater the consequence of fracture.

The architecture of systems thus expanded not only outward in flow, but upward in scale. It built ladders of order, each rung supporting the next, each level enabling survival greater than the sum of its parts.

Section 4: Centers and Peripheries

Every hierarchy drew a map: a center where flows converged, and a periphery where they dispersed. The distinction was not trivial. It defined who commanded, who depended, and how survival was distributed.

In organisms, brains became centers, coordinating signals from distant limbs. Hearts pulsed as hubs, distributing lifeblood to the farthest tissues. The periphery relied on the core, yet the core existed only to serve the periphery. Each needed the other.

In ecosystems, keystone species functioned as centers. Wolves regulated forests. Bees sustained fields. Remove the core, and peripheries unraveled. Yet the smallest peripheries— microbes, fungi, plankton— could destabilize the whole if ignored.

Cultures embodied centers and margins. Capitals gathered wealth, armies, and governance, while provinces depended on their decisions. Markets clustered in cities, pulling resources from hinterlands. Empires rose by centralizing power, and fell when peripheries fractured or withdrew their allegiance.

Technology repeated the pattern. Servers clustered in data hubs, their signals radiating to remote devices. Satellites orbited as silent centers of navigation and communication. Even the internet, imagined as decentralized, revealed central nodes of influence shaping the flows of billions.

Centers and peripheries revealed that architecture was never neutral. It concentrated control, but also responsibility. It created dependence, but also resilience. The periphery was never powerless, and the center was never invincible.

The architecture of systems was always a balance of gravity and orbit, core and edge, each defining the other, each holding the other in place.

Section 5: Stability and Collapse

Architecture promised permanence, but no structure was eternal. Every system carried within it the seeds of endurance and the cracks of collapse.

In organisms, homeostasis balanced stability. Blood pressure regulated, temperature held steady, cells repaired damage. Yet when stresses exceeded thresholds, organs failed, and coherence dissolved into death.

In ecosystems, stability meant balance of species. Predators checked prey, plants replenished soil, climates cycled within limits. But when tipping points were crossed— deforestation, warming seas, extinctions cascading— collapse spread like contagion.

Cultures, too, displayed this duality. Empires stabilized through law and ritual, yet fragility grew with scale. Food shortages, plagues, corruption—each revealed the thinness of stability. Collapse was sudden, but its causes were slow, gathering like pressure in stone.

Technology bore the same pattern. Networks promised resilience through redundancy, but also risked catastrophic failure when central nodes faltered. A power grid might hum for decades, yet fail in seconds. Stability and collapse were two sides of the same design.

Collapse was not always ruin. It cleared the ground for renewal. Forests regrew after fire. Societies reinvented after empire. Technologies replaced the obsolete with the unforeseen. To collapse was to release, to shed structures that could no longer bear their weight.

The architecture of systems was never only about

endurance. It was about rhythm— cycles of stability and collapse, construction and renewal— the grammar of order woven into the inevitability of change.

Section 6: Power and Architecture

Every architecture shaped not only flows, but also authority. To build was to decide, to design was to govern. Power hid in the blueprints of systems.

In the body, the brain directed motion, yet muscles bore the strain. The heart pulsed life, yet relied on vessels to distribute. Power was never absolute. It circulated through the architecture itself, a negotiation of command and dependence.

In cultures, architecture became explicit rule. Walls did not merely defend— they declared ownership. Palaces and temples elevated rulers, casting hierarchy into stone. Roads projected the reach of empire, while prisons embodied authority through confinement of bodies.

Technology extended power's reach. Servers stored the memories of billions. Satellites surveilled entire nations. Algorithms filtered perception, deciding what was visible and what remained unseen. Here, architecture was invisible, yet decisive— power written in code and cable.

Every system revealed the same truth: to control architecture was to control possibility. Who designed the walls, the roads, the networks, decided not only what flowed, but who flourished.

Power and architecture were inseparable. To build was always to choose, and every choice favored some lives over others.

The architecture of systems was thus not only a structure of endurance, but a scaffold of power, a map of who could act, and who was acted upon.

Section 7: Symbols and Monuments

Architecture was never only utility. Walls held back enemies, but they also spoke of sovereignty. Temples sheltered rituals, but they also displayed devotion. Every structure was more than shelter—it was symbol.

Monuments crystallized this truth. Pyramids rose not only as tombs, but as testaments of power. Cathedrals soared to heaven, turning stone into prayer. Towers stretched above cities, announcing ambition in glass and steel. The function of these architectures was inseparable from their meaning.

Symbols gave systems endurance. A palace could be abandoned, but its image lingered. A monument could crumble, yet inspire centuries after.

Architecture became a medium of memory, carrying identity across time.

Cultures fought over symbols as fiercely as over land. To topple a statue was to topple a story. To raise a monument was to inscribe one. Architecture thus became battlefield— a contest not only of armies, but of meaning itself.

Technology reshaped monuments too. Digital skylines now rose in virtual space. Online platforms became symbolic centers, where visibility equaled power. Servers, though windowless, were monuments of another kind— temples of data, repositories of collective memory.

Symbols and monuments revealed that architecture was never mute. It spoke through presence, through endurance, through the stories it carried forward.

To build was to declare. To preserve was to remember. To destroy was to rewrite. Architecture as symbol was thus not only stone and steel, but the shaping of collective identity itself.

Section 8: Networks and Nodes

If monuments embodied permanence, networks embodied motion. Whereas towers and walls stood as fixed symbols, networks spread like webs, linking nodes across distance and time.

In biology, networks bound life together. Neurons fired in patterns of connection. Blood vessels branched to reach every cell. Immune systems signaled across tissues, an invisible web of defense. No node alone was sovereign. Meaning arose in the links.

Ecosystems mirrored this logic. Pollinators linked flowers to fields. Migration routes stitched continents. Nutrient cycles turned decay into growth. Each organism was a node, each relation a thread, and together they wove a fabric of survival.

Cultures discovered networks in trade and communication. Caravans crossed deserts, carrying silk, salt, and story. Shipping lanes braided oceans into economies. Later, telegraphs, phones, and fiber optics linked minds as instantly as bodies. The strength of societies lay not only in centers, but in the resilience of their webs.

Technology radicalized the network. The internet dissolved distance, binding billions into simultaneous exchange. Social graphs mapped relationships, algorithms routed flows, servers pulsed as nodes of light. The architecture of survival was no longer singular or monumental, but distributed and dynamic.

Networks and nodes revealed that endurance could emerge without central command. The web itself was strength, its resilience woven in multiplicity.

To live in networks was to live as part of something larger, an architecture not of stone, but of relation— not fixed in place, but alive in connection.

Section 9: Transparency and Opacity

Architecture did not only shape flows. It shaped what could be seen. Every system carried choices about visibility and concealment— what was transparent, what was opaque, and to whom.

In biology, skins revealed health through color and texture, yet hid inner processes from view. Cells exposed receptors while shielding fragile DNA. Life survived by balancing exposure with protection.

In cultures, transparency became trust. Laws published openly invited accountability. Markets relied on prices visible to all. Yet opacity sustained power too. Secrets guarded states. Mysteries preserved authority in priesthoods. What was hidden could be as influential as what was shown.

Technology magnified the stakes. Surveillance systems watched billions, collecting flows of data unseen by those observed. Encryption offered opacity in return, protecting privacy against intrusion. The architecture of digital life was built as much on visibility as on concealment.

Transparency was never absolute virtue. Opacity was never absolute vice. Too much transparency exposed weakness, too much opacity bred distrust. Survival demanded balance— to reveal enough for coherence, to hide enough for protection.

Transparency and opacity revealed that architecture was not only material. It was epistemic. It decided what systems knew of themselves, and what others could know of them.

To design a system was to design its windows and its walls, its signals and its silences. In every architecture, visibility was power.

Section 10: Adaptability and Rigidity

No architecture was built for eternity. Every structure confronted change— climates shifted, resources waned, pressures mounted. How systems endured depended on their capacity to adapt.

In biology, rigidity meant extinction. Species unable to alter behaviors or bodies vanished when environments changed. Those with flexibility—generalists, shapeshifters, migrants— survived upheavals. Adaptability was resilience written into evolution's code.

In cultures, the same rule held. Rigid empires collapsed when they could not reform. Rituals that never bent lost relevance. By contrast, societies that absorbed novelty, rewrote laws, and welcomed exchange endured beyond crisis.

Technology, too, revealed the balance. Machines designed for one task became obsolete when needs shifted. But modular systems, open protocols, and flexible networks adapted to unforeseen futures. Rigidity guaranteed failure. Adaptability permitted survival.

Yet adaptability was not free. Too much change dissolved coherence. A culture without traditions, a network without standards, a body without stable form— all risked collapse. Rigidity gave identity. Flexibility gave endurance. Each was necessary.

The architecture of systems was therefore judged not only by its strength in the present, but by its capacity to bend with time.

To survive was to strike the subtle balance: rigid enough to cohere, flexible enough to transform, always poised between identity and change.

Section 11: The Aesthetics of Order

Architecture was never only function. It carried an aesthetic dimension, a shaping of form that spoke to senses as much as to needs. Beauty itself became a subtle architecture of survival.

In biology, patterns of symmetry signaled health. Peacock feathers dazzled, but their brightness also marked vitality. Flowers lured pollinators with color and fragrance, binding attraction to survival. Aesthetics was not luxury— it was strategy.

In ecosystems, order appeared as harmony. The rhythm of migration, the balance of predators and prey, the seasonal unfolding of growth— each carried beauty as visible coherence. To witness such order was to glimpse stability.

Cultures transformed aesthetics into art. Temples carved with precision, palaces aligned with stars, cities arranged in geometric grids. Beauty reinforced legitimacy: a ruler who built magnificence claimed power not only through force, but through awe. Even law and ritual were framed in aesthetic order— ceremonies choreographed, texts balanced in rhythm and verse.

Technology absorbed this impulse. Circuits were optimized not only for speed, but for elegance. Interfaces designed for clarity multiplied adoption. Aesthetic simplicity often meant functional strength. What pleased the eye often eased survival.

The aesthetics of order revealed that beauty and survival were intertwined. Patterns that endured were often those that inspired. To build beautifully was to build enduringly, for systems sustained by awe carried not only utility, but devotion.

To design survival was also to design meaning, and meaning often spoke in the language of beauty.

Section 12: Fragility of Structures

However majestic, however intricate, no architecture was without fracture. Every system bore weak points— stresses invisible until tested, fault lines waiting for pressure.

In biology, bones snapped under sudden force. Cells mutated, their delicate codes unraveling into cancer. Brains, exquisitely ordered, were undone by strokes or seizures. Life's architectures balanced resilience with perpetual vulnerability.

Ecosystems revealed the same truth. A keystone species removed collapsed entire food webs. A subtle shift in temperature bleached coral reefs white. Balance sustained for millennia could shatter in decades. What looked stable was often precarious.

Cultures built monuments to permanence, yet history recorded their fragility. Cities buried by earthquakes. Empires dissolved by famine or revolt. Libraries burned, their archives of memory lost in hours. What seemed indestructible proved fragile when tested.

Technology magnified fragility. A single line of faulty code could halt networks of millions. Power grids fell to cascading failures. Aircraft depended on thousands of parts, yet one defect could undo the whole. The greater the scale, the more subtle the vulnerabilities.

The fragility of structures was not only weakness—it was revelation. It reminded us that permanence was illusion, that survival required vigilance, repair, and humility.

Every architecture endured not because it was flawless, but because it was maintained, adapted, and rebuilt in the face of fracture. Fragility was not exception. It was condition.

Section 13: Cycles of Construction and Ruin

Every architecture carried within it the certainty of decay. Stone weathered, wood rotted, steel corroded, code became obsolete. But ruin was not only an end— it was part of a cycle as ancient as life itself.

In biology, cells divided, aged, and died. Apoptosis cleared space for renewal. The body was maintained through constant demolition and repair. Life did not resist ruin; it integrated it.

Ecosystems echoed this rhythm. Forests burned, yet from ash rose fertile soil. Rivers flooded, yet their overflows replenished land. Collapse made renewal possible, weaving destruction into survival.

Cultures, too, cycled through rise and fall. Empires built monuments to permanence, yet each was eventually undone. From the ruins of Rome, new kingdoms emerged. From the ashes of libraries, new knowledge was written. Every collapse opened a clearing for reinvention.

Technology accelerated the rhythm. Devices were built to be replaced. Networks upgraded continuously. What was state-of-the-art today became outdated tomorrow. Here, ruin was not tragedy, but strategy— planned obsolescence as engine of innovation.

The cycles of construction and ruin revealed that survival was not a march toward permanence, but a dance with impermanence. To build was always to build temporarily. To collapse was always to prepare the ground for what came next.

Architecture did not deny ruin. It depended on it. Endurance lay not in resisting collapse, but in learning to rebuild from its inevitable arrival.

Section 14: Memory in Stone and Code

If ruin was inevitable, memory became strategy. Architectures carried not only function, but record— attempts to outlast collapse by embedding meaning in form.

Stone was the first archive. Pyramids preserved dynasties. Inscriptions etched victories into walls. Temples aligned with stars so that centuries later their builders' cosmologies could still be read. Durability became testimony.

But even stone eroded. Wind smoothed faces from statues. Quakes toppled pillars. Time ground permanence into dust. Memory survived only as long as caretakers restored what crumbled.

Cultures sought new substrates. Paper carried scriptures. Printing multiplied voices. Later, magnetic tapes and silicon chips captured sound, vision, calculation. Digital code extended reach further still, allowing memory to proliferate in near-infinite copies.

Yet digital durability was fragile. A single corrupted drive erased entire archives. Formats decayed, servers failed, corporations dissolved. What was infinitely copyable was also infinitely perishable.

Memory in stone and code revealed the paradox of endurance: to preserve meaning was to fight both entropy and obsolescence. No medium was eternal. Each required vigilance, interpretation, translation into new forms.

The architecture of systems thus carried memory forward— not as permanence, but as continuity sustained through care. To remember was to keep rewriting the archive, again and again, across materials and ages.

Section 15: Systems as Environments

At first, architecture sheltered life. Walls, roofs, and enclosures stood apart from the world. But as systems expanded, they ceased to be structures within environments— they became environments themselves.

In biology, the body was more than container. It was habitat for trillions of microbes, an ecosystem within skin. Its systems—circulatory, immune, neural— were environments upon which countless smaller lives depended.

In culture, cities grew until they replaced the landscapes they had been built upon. Markets, schools, and institutions became climates of their own, shaping how people thought, moved, and related. To be born in a city was to be born inside a system as total as weather.

Technology accelerated this engulfing. Networks ceased to be tools and became habitats. Social media was not only platform— it was environment, a place where identity, economy, and politics unfolded. Satellites and sensors now knit the Earth itself into a planetary system, an architecture enclosing the very biosphere.

When systems became environments, escape was no longer possible. One did not live in them occasionally. One lived through them always. The architecture of survival ceased to be backdrop and became atmosphere.

Systems as environments revealed that architecture was not outside of life. It was life's medium. To survive was not to visit a structure, but to inhabit one continuously— a structure so total that it became the world itself.

Section 16: Feedback and Control

When systems grew large enough to become environments, they required new mechanisms to endure. No structure could persist without regulation. Feedback became the hidden architecture of control.

In biology, feedback governed balance. Blood sugar rose, hormones lowered it. Temperature dropped, metabolism increased heat. Every process looped, correcting itself against extremes. Without feedback, life unraveled.

Ecosystems mirrored the rule. Predators checked prey, prey sustained predators. Plants breathed oxygen into animals, animals breathed carbon into plants. Cycles of give and take stabilized environments across millennia.

Cultures embedded feedback in governance and exchange. Courts corrected rulers. Markets adjusted prices. Public opinion swayed policies. Each loop, however imperfect, offered correction to prevent collapse.

Technology multiplied feedback. Thermostats monitored temperature. Algorithms updated themselves based on user behavior. Global networks pulsed with sensors, measuring climate, traffic, finance, disease. Systems now observed themselves, and acted upon their own signals.

Yet feedback was double-edged. A stabilizing loop could become a spiral of collapse. Panic selling crashed markets. Echo chambers amplified extremism. Algorithms optimized themselves into failure. Control could stabilize— or destabilize— depending on how loops were designed.

The architecture of survival was thus not only boundaries, flows, and hierarchies. It was loops of attention and response, self-monitoring circuits that determined whether a system endured or destroyed itself.

Section 17: Fragile Equilibria

Feedback did not guarantee stability. It produced balance, but balance was seldom permanent. Most systems lived on tightropes of adjustment— fragile equilibria where even small disturbances could tip the whole.

In biology, the body balanced water, salt, heat, and

sugar within narrow thresholds. A single fever threatened coherence. A drop in pressure endangered the brain. Life endured only through constant recalibration.

Ecosystems carried the same fragility. A wetland drained, a forest logged, and cycles unraveled. Populations crashed, nutrient flows collapsed, climates shifted into new states. What looked stable was only balance maintained by endless feedback.

Cultures balanced equally on the edge. Justice depended on trust, yet trust could vanish overnight. Economies thrived on confidence, yet panics destroyed them in days. A rumor could destabilize kingdoms, a movement could topple empires. Equilibrium was always temporary.

Technology, too, revealed the delicacy of balance. A network routed millions of signals seamlessly, yet one corrupted packet could ripple into failure. Artificial intelligence tuned itself to optimize outcomes, yet a misaligned signal sent it astray. Every equilibrium rested on vigilance against disruption.

Fragile equilibria revealed that survival was never static. It was dynamic balance, poised between persistence and collapse. The miracle was not permanence, but the ability of systems to recover, recalibrate, and continue forward despite perpetual instability.

Section 18: Systems of Trust

Amid fragile equilibria, one element proved decisive: trust. Without it, feedback faltered, flows broke down, and architectures collapsed. Trust was the hidden architecture binding systems together.

In biology, cells trusted signals. Immune systems recognized self from other. When recognition failed, autoimmunity attacked the body. Even survival at the cellular level was built on fidelity of communication.

In cultures, trust was currency. Markets functioned only when buyers believed sellers, and currencies held value only as long as faith endured. Courts, contracts, and rituals extended this fragile faith, but at root it remained belief in the reliability of others.

Technology inherited this grammar. Networks ran on protocols of trust. Encryption authenticated signals. Users believed systems would not betray their data. When trust was broken— through breach, fraud, or failure— entire architectures unraveled.

Trust was paradoxical. It had to be given in advance, before certainty was possible. Yet it could be destroyed in moments. A thousand transactions built it, a single violation erased it. No architecture could endure without this invisible foundation.

Systems of trust revealed that survival depended not only on walls, flows, or laws, but on bonds unseen. Every empire, every network, every body politic was held together by belief— the conviction that others would act as expected.

To lose trust was to lose coherence. To cultivate it was to sustain survival itself.

Section 19: Collapse of Trust

When trust collapsed, systems unraveled faster than walls could hold or laws could command. Trust was slow to build, but its absence spread like fire.

In biology, cancers ignored cellular agreements. Immune systems turned on the self. Cooperation between tissues dissolved, and the organism failed. Collapse of trust at the cellular level was indistinguishable from disease.

In cultures, corruption corroded institutions. Citizens no longer believed leaders, and obedience eroded. Currencies lost value, markets froze, and revolutions flared. The fall of states was often the fall of faith in them.

Technology revealed the same fragility. When users doubted security, networks emptied. When algorithms proved biased, systems lost legitimacy. Even the most advanced architectures could not survive disbelief. Trust was not optional; it was existential.

The collapse of trust spread contagiously. One breach infected the whole. A scandal in a single office undermined governments. A virus in one machine crippled networks worldwide. Doubt scaled faster than proof.

Yet collapse was not only end. It cleared space for renewal. From distrust rose reform. From failed currencies came new exchange. From broken states emerged constitutions. Collapse forced systems to reimagine their foundations.

The collapse of trust was thus revelation: a reminder that no architecture rested solely on stone or code, but on fragile, invisible bonds— and that every survival was always also a survival of belief.

Section 20: Repair and Renewal

Collapse did not always end the story. Systems, once broken, could be mended. Repair was the slow grammar of survival after failure.

In biology, wounds closed, bones knit, cells regenerated. Even scars carried memory of harm, a record etched into flesh that became part of endurance.

In ecosystems, forests regrew after fire. Wetlands reclaimed fields abandoned. Rivers carved new channels after floods. What seemed ruin became preparation for renewal.

Cultures practiced repair through ritual and reform. Apologies, treaties, and reconciliations stitched trust where it had torn. New laws replaced broken ones. New leaders replaced the discredited. Each act of renewal required acknowledgment of fracture. Denial never repaired. Only honesty restored.

Technology learned repair as well. Patches corrected flaws in code. Networks rebooted after crash. Failures became feedback, guiding stronger designs. Resilience was not in perfection, but in the willingness to rebuild.

Repair and renewal revealed that collapse was not

absolute. It was transition. Every fracture carried the possibility of reformation. Every loss of trust was also the seed of its return.

The architecture of systems was not defined by immunity to failure, but by capacity to mend— to absorb damage, to restore coherence, and to carry memory of collapse as foundation for new strength.

Section 21: Scale of Renewal

Repair looked different depending on the scale at which it worked. A cut on a finger healed in days. A fractured society required centuries. The grammar of renewal shifted with size.

At the level of the body, cells replaced themselves continuously. Bones knit, muscles strengthened with use, immune systems remembered threats. Individual renewal was intimate, measured in heartbeats and scars.

At the level of the community, renewal meant trust restored through dialogue and shared labor. Neighbors rebuilt after storms, families reconciled after quarrels, villages repaired roads and wells. Here, renewal was collective action, visible in cooperation.

Civilizations renewed differently. Collapse of empires gave rise to successor states. Languages evolved from ruins, carrying fragments of memory forward. New religions, constitutions, and markets emerged where old ones failed. Civilizational renewal was not patchwork—it was reinvention, measured in centuries.

Planetary renewal unfolded at scales larger still. Ice ages melted into fertile plains. Forests recolonized barren land. Species radiated after extinctions. Even mass collapse did not erase life—it reset the terms of survival.

The scale of renewal revealed that repair was not one process, but many. Small systems healed quickly. Large ones restructured slowly. Yet all shared the same grammar: endurance through transformation.

To survive was not to resist fracture, but to scale the work of renewal to the size of the wound.

Section 22: Hidden Costs of Renewal

Renewal restored, but it never returned systems to what they had been. Every act of repair left traces— scars, debts, distortions— the hidden costs of endurance.

In biology, healed wounds lost flexibility. Scar

tissue was stronger, yet less supple. Bones reknit with density, but risked stiffness. The body endured, but at the price of altered form.

In ecosystems, forests that regrew after fire did not always match the diversity of what had burned. Species disappeared forever, soil thinned, cycles shortened. Renewal preserved survival, but often diminished abundance.

Cultures, too, bore costs. Reforms emerged from collapse, but trauma lingered in memory. Revolutions wrote new constitutions, yet often carried seeds of old divisions. Debt from rebuilding strained generations yet unborn. Renewal was never free— its price was carried forward.

Technology displayed the same pattern. Software patched flaws, but complexity accumulated. Each layer of repair increased fragility. Systems worked, but under weight of hidden vulnerabilities.

The hidden costs of renewal revealed that endurance was not restoration of purity, but layering of compromise. Every scar was a story. Every repair a negotiation. Survival itself was architecture built on residue.

To endure was to accept that nothing whole remains unchanged. Renewal always alters, and what persists is never what once was— but what could be sustained.

Section 23: Renewal as Reinvention

Renewal was never simple repair. It was reinvention. Systems did not return unchanged to their earlier state. They emerged altered, recast by the very fractures they sought to mend.

In biology, cells replaced themselves with slight variations. DNA copied, but mutations slipped through. Each repair altered the lineage, making life a history of reinventions disguised as continuity.

In ecosystems, collapse opened niches. Species adapted, migrated, or transformed. A ruined forest returned not as replica, but as new ecology. Renewal reshaped what survival meant.

Cultures reinvented after crisis. The fall of Rome gave rise to medieval Europe. The Enlightenment followed wars of faith. Constitutions were drafted from ashes of monarchy. Renewal rewrote identity, turning loss into foundation for novelty.

Technology embraced reinvention outright. Systems were not only patched— they were upgraded. Failures became blueprints for redesign. Collapse was feedback driving the next generation of invention.

Renewal as reinvention revealed that survival was not continuity of the same, but transformation into the new. Every scar carried change. Every collapse seeded novelty. To endure was not to preserve identity intact, but to let identity evolve.

The architecture of systems was thus a workshop of reinvention— where even failure became material, and where every act of repair was also an act of creation.

Section 24: Renewal and Memory

Reinvention did not erase history. Even as systems transformed, they carried traces of what had been. Renewal required memory, for without memory, coherence dissolved.

In biology, healed tissue bore scars. Genomes carried mutations but also preserved ancestral codes. The new body was never wholly new. It stood on the sediment of its past, remembered in every cell.

Ecosystems regenerated by drawing on memory of soil and seed. Forests regrew with species adapted to fire. Rivers reclaimed their beds after floods. Memory guided resilience, anchoring reinvention in continuity.

Cultures rebuilt with archives intact. Myths retold gave context to change. Histories warned of dangers. Libraries, even half-burned, seeded futures of thought. Revolutions declared novelty, yet drew legitimacy from stories of what came before.

Technology remembered as well. Legacy code persisted in new systems. Old protocols underpinned modern networks. Even when reinvented, systems carried fossils of design, layers of memory sedimented into function.

Renewal and memory revealed that survival was not blank slate. It was palimpsest— a rewriting that never erased fully, a layering of new meaning upon remnants of the old.

To endure was to remember enough to remain coherent, while forgetting enough to allow transformation. Memory held continuity. Reinvention carried it forward.

Section 25: Renewal as Collective Work

No system renewed itself alone. Repair required cooperation, reinvention demanded collaboration. Survival was collective labor woven across scales.

In biology, cells coordinated in healing. Platelets

clotted blood, fibroblasts laid scaffolds, immune cells cleared debris. No single cell repaired the wound. The body restored itself through concert of many parts.

In ecosystems, renewal followed symbiosis. Fungi decomposed, insects pollinated, plants recolonized bare soil. Life returned not through solitary species, but through networks of interdependence.

Cultures repaired themselves in shared effort. Villages rebuilt walls together. Communities restored trust through ritual, dialogue, and care. Nations recovered from war only when millions labored to feed, govern, and reconcile. Renewal was always a work of the many.

Technology embodied the same principle. Open-source projects evolved through collective coding. Global networks stabilized through cooperation of protocols. Even the internet itself was renewal as collective work, maintained by invisible hands across continents.

Renewal as collective work revealed that survival was not authored by isolated actors. It was ensemble performance, a chorus of agents holding systems together.

To endure was to join the work— to repair not only

for oneself, but with and for others, so that the architecture of survival could extend beyond any single life.

Section 26: Renewal and Imagination

Repair stitched what was broken, but imagination envisioned what had not yet been. Renewal depended on both: the work of hands and the vision of minds.

In biology, healing was guided by form. Cells knew not only to close a wound, but to restore shape, following a memory of structure. Imagination was nature's blueprint, encoded in the body's design.

In ecosystems, renewal required anticipation. Seeds lay dormant, awaiting rains not yet fallen. Animal migrations traced paths to lands they had not yet seen. Life renewed itself through foresight as much as feedback.

Cultures wove imagination into every act of rebuilding. Cities destroyed by war were remade with visions of peace or grandeur. Constitutions drafted after collapse imagined freedoms not yet lived. Art, myth, and philosophy stretched renewal beyond necessity into possibility.

Technology thrived on imagination. Every repair was an opportunity for redesign. Failures became prototypes. Systems broken today seeded innovations tomorrow. Imagination made renewal not only restoration, but expansion.

Renewal and imagination revealed that survival was not only endurance, but creativity. To mend what was broken was to picture what could be. To endure was to dream repairs that exceeded necessity and opened futures.

Imagination turned fracture into foundation. It transformed collapse into the canvas of what comes next.

Section 27: Renewal and Time

Renewal was never instant. Its rhythm unfolded in time, and the tempo of repair determined the destiny of systems.

In biology, some wounds closed within hours, while others lingered for years. Cells hurried to knit skin, yet nerves regenerated slowly. The pace of renewal defined resilience, measured in clocks of flesh.

In ecosystems, renewal could take centuries. A forest felled might not return for generations. Soil rebuilt slowly, species recolonized cautiously. The tempo of repair was geologic compared to human life.

Cultures renewed across decades. Institutions eroded gradually, then reformed through upheaval. Revolutions flared quickly, but their reforms stabilized only with time and patience. Societies mended not in moments, but across lifetimes.

Technology accelerated renewal. Software patched overnight, networks rebooted in seconds, devices replaced in months. Here, time compressed—renewal became constant cycle, measured in updates, not in generations.

Renewal and time revealed that survival carried many tempos. Some wounds demanded speed, others endurance. Too fast, and coherence shattered. Too slow, and collapse overtook repair.

The architecture of systems was thus not only about repair, but about timing— knowing when to act swiftly, when to wait patiently, and how to live within rhythms larger than one's own span.

Section 28: Renewal and Vulnerability

To repair was to expose. Every act of renewal opened wounds before they closed, revealed weaknesses before they healed. Vulnerability was not failure— it was condition of restoration.

In biology, healing left bodies open to infection. A scab defended, yet remained fragile. Immune systems labored hardest when defenses were weakest. The path to wholeness passed through exposure.

In ecosystems, renewal invited invasion. Burned forests, though fertile with ash, were vulnerable to erosion or weeds. Rivers reshaping after floods could carve away stability. Regrowth was never guaranteed.

In cultures, renewal exposed political fractures. Revolutions cleared ground, but also opened struggles for power. Reforms invited backlash. Periods of rebuilding were also moments of greatest instability. The chance for renewal was also the chance for exploitation.

Technology revealed the same risk. Software patched against flaws might introduce new ones. Upgrades solved vulnerabilities while exposing others. Every renewal carried the possibility of deeper fragility.

Renewal and vulnerability revealed that survival was never without risk. To rebuild was to open oneself again to uncertainty and danger.

The strength of systems lay not in avoiding exposure, but in withstanding it— in passing through the weakness of renewal and emerging stronger on the other side.

Section 29: Renewal as Design

Not all renewal was accidental. Some systems prepared for collapse, anticipating fracture and building repair into their form. Renewal became a matter of design.

In biology, redundancy was architecture. Two kidneys sustained the body so one could fail. Paired lungs, paired eyes, backup pathways in the brain—life embedded renewal in its structure, ensuring survival through loss.

Ecosystems designed resilience through diversity. Multiple pollinators ensured flowering. Redundant species stabilized food webs. If one link faltered, others carried the load. Diversity was not excess, but insurance against ruin.

Cultures, too, built renewal into institutions.

Checks and balances dispersed power, so collapse of one branch did not undo the whole. Succession laws ensured continuity. Rituals marked renewal of seasons, embedding rhythm into tradition.

Technology followed the same logic. Distributed networks rerouted traffic when nodes failed. Cloud storage preserved copies beyond local collapse. Some systems were designed to degrade gracefully, never shattering all at once.

Renewal as design revealed that fragility could be anticipated, and recovery engineered. To design for endurance was not to resist failure absolutely, but to absorb it, bend with it, and rise rebuilt.

Survival's architecture was not permanence, but the foresight to make collapse part of the plan.

Section 30: Renewal and Creativity

Renewal was not only recovery. It was creation. The very act of rebuilding invited invention, turning necessity into possibility.

In biology, mutations slipped into repair. Most were silent, some destructive, but a few reshaped lineages forever. The creativity of life was hidden in its errors, its willingness to rebuild differently.

In ecosystems, collapse cleared space for novel forms. Plants colonized fresh ash, species adapted to new climates, invasions rearranged webs of relation. Every disturbance was also experiment.

Cultures transformed breakdown into art. Wars ended, and poets wrote new epics. Ruined cities inspired new architectures. Philosophies emerged from the ashes of disillusionment. Renewal was as much imaginative as structural, woven from dreams of what could replace what had been lost.

Technology formalized this cycle. Failures seeded prototypes, patches evolved into redesigns. Obsolescence bred innovation, as each discarded model made space for invention. Renewal became deliberate creativity, a system of perpetual reinvention.

Renewal and creativity revealed that survival thrived not in preservation, but in transformation. To endure was to create again, to find beauty in fracture, and to let collapse become the canvas for novelty.

The architecture of systems was never only defense against loss. It was a studio of invention, where each wound opened a door to the unexpected.

Section 31: Renewal and Justice

Every collapse left questions not only of structure, but of fairness. Renewal was never neutral. It distributed burdens and benefits, deciding who bore the cost of ruin and who inherited the fruit of repair.

In biology, healing favored survival of the whole, yet not all tissues healed equally. Some wounds restored quickly, others left lasting damage. The body survived, but not every part returned the same.

In ecosystems, renewal privileged certain species. Fast-growing plants shaded slower ones. Invasive arrivals claimed new niches. The forest returned, but not as it was— balance tilted, and some forms were excluded.

In cultures, justice defined renewal. After wars, reparations and treaties decided futures of nations. After oppression, movements demanded recognition, restitution, inclusion. Renewal without justice preserved inequity, layering old wounds beneath new structures.

Technology repeated the dilemma. Access to repair and upgrade was uneven. Some communities rebuilt swiftly, others languished. Even innovation carried bias— its benefits clustered where wealth and power already lay.

Renewal and justice revealed that endurance required more than functionality. It demanded fairness. To rebuild without equity was to plant seeds of collapse within the new order.

True renewal did not simply restore. It redistributed. It confronted wounds honestly, so that survival meant not only continuation, but transformation into something more just than before.

Section 32: Renewal and Vision

Repair alone was not enough. Without vision, renewal risked becoming mere patchwork— a cycle of short-term fixes that left deeper fractures untouched. Endurance required foresight.

In biology, healing followed patterns encoded in the body. Cells rebuilt not randomly, but according to form. Vision was memory of shape, a map inscribed in tissue that guided repair beyond instinct.

In ecosystems, renewal stretched over centuries. Forests regrew not for one season, but for generations. Seeds adapted to fire or frost were strategies for futures unseen by their planters. Vision in nature was patience.

Cultures rebuilt most lastingly when guided by vision. Revolutions faltered if they sought only revenge. Nations endured when constitutions looked forward, imagining freedoms and rights beyond immediate grievance. Vision turned survival into legacy.

Technology revealed the same lesson. Patching flaws without redesign invited future collapse. But systems planned with scalability, redundancy, and openness endured longer. Innovation without vision was fragile. Vision without action was empty. Renewal demanded both.

Renewal and vision revealed that survival was not repair of the past, but preparation for the future. To endure was not only to mend, but to imagine—to see beyond urgency into the shape of what might come.

The architecture of systems stood firm not by accident, but by design guided with foresight. Without vision, renewal decayed into repetition. With vision, it became creation.

Section 33: Renewal and Collective Memory

Vision carried systems forward, but memory anchored them. Without memory, renewal risked repeating collapse. Without vision, it risked stagnation. Endurance required both.

In biology, immune systems remembered pathogens. Each infection left a trace, a record guiding future defense. The body renewed not blindly, but with archives of survival etched into cells.

In ecosystems, memory lived in soil and seed. Forests regrew in patterns set by centuries of fire and flood. Rivers carved valleys that shaped every future flow. Nature renewed with memory, its past encoded in landscape.

Cultures cultivated collective memory. Rituals retold origins. Monuments marked sacrifice. Histories preserved warning and pride. When memory fractured, renewal faltered— ignorance of past injustice sowed new divisions, denial of past mistakes invited repetition.

Technology expanded memory's reach. Archives stored data for centuries, digital traces multiplied endlessly. But excess memory risked paralysis. Too much preservation blurred clarity with noise. Renewal required discernment: what to carry forward, what to let go.

Renewal and collective memory revealed that survival was both inheritance and choice. To endure meant learning from wounds already lived, yet daring to build beyond them.

The architecture of systems was always double—facing backward to remember, facing forward to imagine. Between memory and vision, renewal found its path.

Section 34: Renewal and Responsibility

Renewal was never neutral. To rebuild was to decide, to shape futures for others as well as oneself. Every act of repair was also an act of responsibility.

In biology, cells divided faithfully, but errors rippled outward. A mutation unchecked could endanger the whole body. Responsibility in life was written as fidelity— accuracy in the smallest acts to sustain the larger system.

In ecosystems, species bore responsibility through interdependence. Pollinators sustained plants, plants sustained grazers, grazers sustained predators. When one species overreached, balance fractured. Responsibility was survival shared across roles.

In cultures, renewal demanded accountability.

Leaders rebuilding after collapse chose who would benefit, who would bear cost, who would be remembered, who forgotten. Responsibility could not be escaped— it was embedded in every decision of reconstruction.

Technology carried responsibility too. Designers shaped systems that governed trust, protected privacy, or exploited it. Repair of networks and machines was never purely technical— it determined who was empowered, and who was exposed.

Renewal and responsibility revealed that survival was not merely continuation. It was stewardship. To endure meant to carry the weight of others, to rebuild with awareness that every choice inscribed a future someone else must live.

Responsibility was the unseen scaffold of lasting renewal. Without it, repair was exploitation. With it, renewal became justice.

Section 35: Renewal and Hope

Responsibility carried the weight of survival, but hope carried its momentum. Without hope, renewal collapsed into obligation. With hope, it became possibility.

In biology, hope was not sentiment but instinct.

Cells dividing after injury persisted without guarantee. The body repaired because continuation was written within it. Hope was the drive to attempt renewal despite uncertainty.

In ecosystems, hope was visible in resilience. Green shoots pierced charred soil after fire. Birdsong returned to silent valleys. Life reached forward even when ruin seemed final. Hope was survival's rhythm made visible.

In cultures, hope animated recovery. Communities rebuilt after storms, not knowing if storms would return. Nations drafted new laws while memories of tyranny lingered. Artists painted futures while living in rubble. Hope turned despair into labor, into shared commitment to begin again.

Technology embodied hope in design. Engineers patched systems believing stability could be restored. Inventors created as though collapse was temporary. Every upgrade, every innovation, was an act of faith in continuity.

Renewal and hope revealed that survival required more than structure or foresight. It required belief that tomorrow was worth building. Hope was not naivety— it was discipline, the courage to repair when outcomes could not be promised.

Hope was the quiet architecture beneath every act of endurance. It bound memory and vision into the willingness to try again.

Section 36: Renewal and Generations

Renewal was never confined to a single lifetime. Its work stretched across generations, each inheriting fractures and repairs from those before, each leaving traces for those yet to come.

In biology, renewal was lineage. DNA carried scars of mutation, but also adaptations that endured. A wound in one body healed, but the deeper renewal was in the survival of offspring. Life continued through inheritance, repair passed forward as code.

In ecosystems, generational renewal was visible in cycles. Trees shed seeds that grew into forests long after the parent trunk had fallen. Rivers carved valleys that guided migrations for millennia. Each generation of life built frameworks for the next.

Cultures marked renewal in the transfer of memory and duty. Elders taught stories, parents rebuilt homes, teachers passed knowledge. Collapse became lesson, renewal became inheritance. Even revolutions, though claiming novelty, were shaped by debts to the generations they replaced.

Technology accelerated inheritance. Knowledge stored in archives, designs passed through networks, inventions seeded futures unimaginable to their creators. Each generation of devices was scaffold for the next— renewal at mechanical speed.

Renewal and generations revealed that survival was never solitary or present-tense. It was relay. Each life a carrier, each system a bridge. To endure was to hand forward structures of possibility, so the next could build further than the last.

Renewal was thus not only for the living. It was for those who waited in silence, unborn inheritors of what we choose to repair and create.

Section 37: Renewal and Identity

Every act of renewal reshaped not only structures, but selves. Identity was never static— it was rewritten each time systems broke and healed.

In biology, cells replaced themselves continually. The body you inhabited at birth was not the body you carried decades later. Renewal did not preserve identity intact. It reconstituted it, assembling continuity from change.

In ecosystems, renewal redrew belonging. Species migrated, adapted, or vanished. Forests reemerged with new cast of actors. Identity of place shifted with every cycle of collapse and return.

In cultures, identity was the deepest site of renewal. Colonies reclaimed independence. Communities redefined themselves after oppression. Generations born into new laws lived as if older injustices had never been permissible. Renewal reshaped who they were, not only what they had.

Technology magnified this rewriting. Digital repair altered how people presented themselves. Archives of memory were curated, profiles edited, histories erased or reinvented. Renewal became identity's workshop, shaping the story of the self as much as the structure of society.

Renewal and identity revealed that endurance was not only structural, but existential. To repair was to redefine. To survive was to accept that who we are is built not on permanence, but on the capacity to be remade.

Identity itself was an architecture of renewal— a bridge carried forward from fracture into coherence, from collapse into becoming.

Section 38: Closing Reflection on Renewal

From wounds that healed in flesh to civilizations rebuilt from ruin, renewal revealed itself as the grammar of endurance. Not permanence, not purity, but the weaving of fracture and repair into continuity.

Biology taught renewal as regeneration. Ecosystems framed it as cycles. Cultures inscribed it in rituals, laws, and revolutions. Technology accelerated it into perpetual iteration. Every domain confirmed that survival was not the absence of collapse, but the capacity to rise within it.

Renewal was never restoration alone. It was reinvention. Each scar carried change. Each collapse seeded novelty. Every repair rewrote identity, and every generation inherited both wounds and their mending.

To endure was not to return unchanged. It was to remember, to reimagine, and to build again—different, yet continuous. The past survived in memory, the future emerged in vision, and the present labored to stitch them together.

Chapter Five closes here. We have traced

architecture from boundaries to flows, hierarchies, collapse, and renewal. The lesson is clear: systems endure not because they are unbroken, but because they can be broken and still rebuilt.

In Chapter Six, we will turn from renewal to expansion— to the architectures that do not only persist, but scale outward, engulfing entire worlds of relation. Where renewal kept systems alive, expansion carried them forward, multiplying their reach until survival itself became empire.

Chapter Six — The Expansion of Systems

— * —

Thesis

Survival did not stop at endurance. Once systems renewed themselves, they sought to extend. Expansion was the next architecture— the drive to scale outward, to multiply boundaries, to project flows and power beyond their origin.

Expansion was not accident. It was momentum. Cells divided into tissues, tissues into organisms. Villages became cities, cities became empires. Small networks multiplied into global webs of connection.

To expand was to translate resilience into influence, to carry what survived into greater reach. But expansion was double-edged: it offered abundance and connectivity, yet carried risk of overreach, domination, and collapse.

Here, we trace how systems expand— biological, ecological, cultural, and technological— and how their growth reshapes the very conditions of survival.

Section 1: The Impulse to Grow

Life carried within it an irrepressible urge to grow. Cells multiplied, branches stretched toward light, rivers carved deeper valleys. Expansion was not luxury, but instinct— an extension of survival itself.

In biology, growth meant replication. From single cells to colonies, from embryos to bodies, multiplication ensured continuity. The impulse to grow was written into the very code of life.

In ecosystems, growth meant spread. Forests advanced across plains, coral built reefs from fragments, species colonized new niches. Growth was adaptation extended outward, reshaping environments as it reached.

In cultures, growth meant increase in number and influence. Families extended into clans, clans into tribes, tribes into nations. Markets widened their reach, religions expanded through conversion, languages spread across continents. Expansion gave endurance the strength of scale.

Technology amplified this impulse. Circuits miniaturized even as networks multiplied. Each invention begot more, each connection spawned others. Systems expanded not only in size, but in speed and density.

The impulse to grow revealed that expansion was not deviation from survival. It was survival's extension— the will of systems to exceed themselves.

Section 2: Boundaries Extended

Expansion began not in conquest, but in the quiet act of stretching lines. Every system that endured pressed against its edges, testing where inside ended and outside began.

In biology, cells grew until membranes touched, then pushed outward into tissues. Skin thickened, bones lengthened, organs unfolded into space previously unclaimed. Boundaries extended so that survival could enlarge its reach.

In ecosystems, boundaries shifted with migration. Wolves expanded territories, plants spread through wind-blown seed, invasive species crossed oceans to redraw ecological borders. Expansion remade maps of relation, displacing some, absorbing others.

Cultures extended boundaries in visible ways. Walls were pushed outward, fields annexed, frontiers declared. Nations marked rivers and mountains, then erased them with treaties. Colonization carried this logic further, stretching boundaries until one world absorbed another.

Technology repeated the pattern. Networks added new nodes, signals leapt across satellites, markets reached into households previously untouched. Every expansion extended what was counted as inside the system, shrinking the realm of outside.

Boundaries extended revealed that expansion was not merely addition. It was redefinition. To expand was to declare that what was once beyond now belonged within.

Every system redrew the map of itself— not only to survive, but to claim a larger share of existence.

Section 3: Scaling Networks

Expansion was not only a matter of space. It was also a matter of relation. Systems grew stronger not merely by stretching outward, but by weaving more threads between their parts.

In biology, neurons extended axons, branching into webs of connection. Blood vessels multiplied to feed growing tissues. Strength came not from size alone, but from density of links.

In ecosystems, networks scaled through interdependence. Bees linked orchards across valleys. Migrating herds stitched landscapes together. Currents carried nutrients from ocean depths to coasts. Expansion unfolded as connectivity, binding distant places into one web.

In cultures, roads carved paths across continents. Caravans joined deserts to seas. Trade routes like the Silk Road turned strangers into partners. Later, railroads and telegraphs compacted space into hours and minutes. Expansion was measured in how fast connections could be made.

Technology accelerated this logic. Networks multiplied nodes at exponential rates. Satellites orbited above, fiber optics pulsed below. Billions of devices interlaced into a single system, so vast it appeared invisible, yet intimate in every life.

Scaling networks revealed that expansion was not simply spread. It was thickening— increasing ties until distance dissolved, until survival depended not on isolation, but on interconnection.

To expand was to weave. To weave was to endure.

Section 4: Expansion and Competition

Expansion did not unfold in empty space. Every system's growth met the growth of another. Boundaries overlapped, networks collided, and survival became contest.

In biology, species competed for territory, for food, for mates. Roots tangled underground, branches fought for light. Predators pressed against prey, prey evolved defenses in return. Expansion sharpened into struggle, a perpetual trial of adaptation.

In ecosystems, competition shaped balance. One species' gain was another's decline. Invasive arrivals often flourished by outcompeting natives. Expansion here was never neutral— it reordered webs of relation through rivalry.

Cultures embodied the same law. Cities warred for resources. Nations clashed at borders. Empires expanded only by subduing others. Markets, too, thrived on competition, turning expansion into a struggle for dominance of trade and influence.

Technology magnified rivalry. Corporations scaled networks to capture markets. Algorithms competed for attention. Even global infrastructures— energy grids, satellite constellations, communication networks— contested space and spectrum.

Expansion and competition revealed that growth was never pure increase. It was redistribution. One system's expansion often meant another's contraction. Survival in expansion required not only growth, but strategy against rivals.

To expand was to confront others. To survive was to learn how to share, resist, or dominate in the crowded architecture of life.

Section 5: Expansion and Cooperation

Competition was not the only path. Systems discovered that survival could be multiplied through alliance. Expansion could proceed not by conflict, but by cooperation.

In biology, cells joined in symbiosis. Mitochondria once free-living became partners within cells, powering life through union. Lichens bound fungi and algae into single thriving organisms. Here, expansion was achieved not by conquest, but by collaboration.

In ecosystems, mutualism drove abundance. Pollinators spread flowers, flowers fed pollinators. Cleaner fish groomed predators, predators spared them in return. Entire biomes flourished through cooperative ties that scaled beyond rivalry.

Cultures practiced cooperation through trade, treaties, and federation. Villages exchanged surplus. Nations formed alliances. Religions spread not only by force, but by conversion, hospitality, and exchange. Cooperation expanded influence without exhausting resources in war.

Technology embodied the principle deeply. Networks interoperated through standards. Open-source projects thrived on shared labor. Global infrastructures— from climate monitoring to space exploration— required nations to cooperate beyond rivalry.

Expansion and cooperation revealed that growth was not always zero-sum. When systems aligned, they scaled further together than apart.

To expand was not only to dominate. It was also to connect, to multiply strength through shared survival.

Section 6: Expansion and Overreach

But expansion— unexamined— turns into hunger.

The same drive that sparked innovation can grow beyond its frame. What once ensured survival may now threaten it.

Forests cleared beyond regrowth. Fisheries harvested beyond renewal. Cities spread into deserts, into floodplains, into fault lines.

In biology, cancer is a cell that forgets its role— multiplying for growth's sake, until it consumes its host.

In society, empires collapse from the weight of overextension. Supply lines stretch too far. Governance fragments. Resources dwindle. Revolts ignite.

Even knowledge— when grown without integration— creates silos, echo chambers, systems so complex they fail under their own weight.

Overreach is not always visible. It begins subtly: a tolerance here, an externality there. But patterns compound. Feedbacks spiral. And collapse comes not by sabotage, but by saturation.

Every expansion creates a new edge— a place where the system must balance, or break.

Section 7: Expansion and Collapse

Every system stretched without feedback meets a wall.

There is no empire immune. No species exempt. No algorithm above entropy.

Collapse is not failure. It is return. Return to equilibrium, to constraints, to structure.

In ecology, a population that overshoots consumes its base and crashes.

In finance, a market that inflates must correct or implode.

In civilizations, growth without redistribution hollows the center, ignites the margins. Collapse begins not with loud destruction but with silent withdrawal.

The soil goes fallow. The workers disappear. The rituals lose meaning. The walls no longer hold.

Collapse is structure enforcing its terms. It is the grammar of reality reasserting syntax on a story written too fast.

It is not vengeance. It is calibration.

Section 8: The Memory of Collapse

Collapse leaves fingerprints. Even when erased, its imprint shapes the survivors.

The structure remembers what the story forgets. Scars become instructions. Ash becomes nutrient. Failure becomes constraint.

A forest that burns regrows with fire-adapted bark. A lineage that nearly ends mutates, adapts, evolves a warning.

A civilization that falls writes its laws tighter. Builds lower. Fears deeper.

Collapse becomes a feedback loop embedded in the blueprint.

Genes carry it. Traditions encode it. Architecture reflects it.

Collapse is not just a past event. It is a present structure. A silent instructor.

It doesn't teach with lectures. It teaches by removal. By what no longer stands.

The absence becomes the clearest pattern. What failed defines what survives.

Section 9: The Architecture of Restraint

What looks like caution is often the fossil of collapse.

What looks like tradition may be the memory of one too many fires.

Every restraint has a lineage. No society fears randomly. No boundary appears without ancestral wounds beneath it.

The bridge is not overbuilt— it is built by those who lost people to collapse.

The wall is not merely symbolic— it is the echo of forgotten floods.

What we call overdesign is often precise historical scar.

Cells develop checkpoints after too many divisions led to ruin. Brains build inhibition after too many risks end lives. Algorithms insert safeguards after cascading failures.

None of it is arbitrary. All of it is memory.

Restraint is not weakness. It is survival's geometry. It is pattern shaped by prior pain. It is collapse, inverted into structure.

Section 10: Constraint as Intelligence

The mind is not free because it thinks. It thinks because it is not free.

Freedom does not birth intelligence. Constraint does.

A river does not wander everywhere. It is carved by the valleys it cannot escape. Its power grows not by directionless flow but by the narrowing of choice.

The neuron does not spark because it may do anything— but because it may do one thing in the right place.

Language does not emerge from infinity. It comes from sounds that could be made by a throat. Then sounds that could be shared. Then those that could be remembered.

Every intelligence is compression. Every learning system is shaped by loss. Every insight is a constraint that finally lined up with a path that held.

You understand not because you saw everything— but because something ruled out all the rest.

To be intelligent is not to be limitless. It is to be exquisitely limited in just the right ways.

Section 11: The Blueprint is Not an Option

A blueprint does not offer choices. It offers certainty. Its lines are not suggestions— they are sentences already passed.

DNA does not ask what form to take. It unspools. It folds. It repeats what was written long before the organism awoke.

Cities do not grow by whim. Roads follow earlier paths, markets rise where trade already gathered, walls extend from forgotten boundaries. The blueprint persists long after the builders are dust.

Even thought obeys it. Ideas arise not from anywhere, but from what language and memory permit. What seems like creativity is blueprint's echo— a new sentence but not a new alphabet.

The blueprint is indifferent. It does not bend to preference or plea. Its structure carries forward, unmoved by what believes itself to be the author.

What you call possibility is already narrowed. What you call choice is already traced.

The blueprint is not an option. It is the stage, the script, the scaffolding of existence.

You walk lines that were inked before you arrived. The shape was waiting. The rest is only performance.

Section 12: Expansion Through Inheritance

What grows does not grow freely. It extends what it received. From vines to values, from dendrites to doctrine, expansion is not liberation— it is replication with deviation measured in tolerated error.

The oak carries its acorn's limits. Its canopy is vast, but not directionless. Every branch obeys an inherited architecture.

So too in civilization. Empires rise not by invention but by iteration. They extend laws carried forward like root structures beneath new soil.

Languages evolve, but within scaffolds. New words do not rupture grammar— they inhabit it. They are mutation, not exodus.

Even your values— compassion, justice, beauty— are echoes of inherited forms, whispers from ancestral codes passed through education, emotion, environment.

Expansion is rarely escape. It is inheritance in disguise. The novel is the remembered, with its face turned slightly.

You did not build the ground you walk on. You expanded upon it. And what seems like departure is merely a new hallway in an ancient structure.

To inherit is to grow— but only where the blueprint permits. Even freedom must pass through the gates of form.

Section 13: Instructions in Disguise

What appears spontaneous is structured. Even improvisation follows rules— unspoken, but inherited. The jazz musician hears freedom, but fingers trace scales taught in silence.

The poet breaks form yet begins in form. Blank verse is never blank; it is echo.

We call it "talent," but it is transcription. Signals passing through fingers trained by repetition, reinforced by culture, shaped by scarcity, reward, approval.

The dancer does not invent motion. They retrieve it from anatomy, from tradition, from gravity.

This is the hidden engine: an instruction set coded in constraint, masquerading as inspiration.

You may decorate the path, but you do not lay it. The steps you take were carved before you. To walk them is not to choose them.

Even deviation is rehearsed. Even rebellion is rehearsed. A revolution that spreads does so because it obeys propagation laws older than the cause itself.

So when you feel the spark— pause. Not to question its fire, but to recognize the flint. That strike was shaped before you struck.

Section 14: The Illusion of Original Motion

A comet burns bright— but not because it chose to. Its path is prewritten, etched in orbital debt, its brilliance a byproduct of exposure, not invention.

So too with us. The moment you think you've moved in a new direction, pause. That turn was a tunnel. You only saw light because the structure let you.

To create is to reroute— never to originate. What we call "new" is seldom uncaused. The conditions always exist before the expression.

You think you shifted course. But what shifted was your position within a pre-shaped flow. A tributary didn't decide to join the river. It simply reached the edge.

Original motion, like originality itself, is a ghost costume draped on causality. Beneath the disguise: gradient descent. Feedback loops. Adaptive systems tuned by constraint.

This is not poetry. It is architecture wearing metaphor.

Every invention— a necessary next step of material logic. The lever. The loom. The chip. They arrive not by genius but by pressure. When enough load builds, the outcome appears inevitable— then gets named "breakthrough."

But the true force was always silent: Structure bending toward complexity, not applause.

Section 15: The Applause Was Premature

The audience rose— not for the structure, but for the spark that stood at the edge and lit what was already soaked in fuel.

Praise arrived like lightning before thunder— instant, visible, wrong. The system had long prepared what they mistook for spontaneity.

He was called brilliant because the doors opened when he pushed. No one praised the hinges. No one clapped for the unseen pressure that built behind the wall until a single finger looked divine.

A design reaches the world, and they cheer the hands that drafted it. But the materials, the equations, the unsolved needs, and the surrounding constraints— those went unnoticed.

Every ovation was premature. The moment was not a miracle, but a mechanical result. Like water pressing against a dam— it was not the crack that deserved the ceremony.

Even Pamela's cell factories, her modular proteins, her genetic scaffolds— they came not from thin air but from the accumulated dust of every model, every failed scaffold, every constraint observed and then re-expressed.

The real miracle was not that the idea emerged— but that it had nowhere else to go. It was not born. It was cornered. And we clapped because we didn't see the maze.

Section 16: The Maze Was Always There

They thought they wandered freely— but the walls were already built. Left turns were finite. Dead ends were intentional. And the destination? It had been waiting.

Pamela did not invent the future— she reached the only exit the system allowed. Her genius was not a spark, but a submission. A perfect fit for a maze that had only one open path.

The idea was praised, but the terrain was ignored. What emerged was not miraculous, but architectural. The walls guided her hands. The constraints carved the logic. Each advancement was an inevitability with excellent timing.

Innovation, they say, but what they meant was

discovery of a corridor already narrowing since the first cell divided.

The maze was biochemical. It was economic. It was sociotechnical. It was the convergence of failed attempts, distant whispers, discarded maps, and misremembered blueprints.

Her "breakthrough" was just the moment the walls gave up trying to resist what they had already shaped.

Section 17: Genius Is Just a Perfect Fit

What they call genius is often nothing more than a shape that fits cleanly into a pre-cut void.

Pamela was not the first to try. She was the first whose structure matched the missing piece.

No force was required. The system welcomed her. Each molecule, each idea, each lab bench, had been angling for her entrance.

She walked in and was called visionary. But vision had nothing to do with it.

The world had carved a niche so exact that only one outline could slide in without friction.

Even the resistance she faced was calculated friction. It shaped her edge, refined her approach, polished the fit.

And when the connection clicked, the machine called it genius. But the machine was just closing a circuit. It needed her not because of what she was—but because of what she matched.

A pattern needs no permission to complete itself.

Section 18: Failures That Made the Fit Possible

Every rejection left a dent.
Every funding cut chiseled the edge of what was no longer allowed.
Every abandoned theory hollowed out a chamber waiting for the next attempt.

By the time Pamela arrived, the shape of failure was already familiar. She did not overcome it—she completed it.

Failed attempts at reprogramming cells. Missed targets in metabolic engineering. Botched delivery systems. Years of trial and no headline.
Each was an architect of her inevitability.

The world forgets that structure is not just built— it is unbuilt. It is defined by what no longer works. By every plan that was too early, too rigid, too proud.

She arrived not to triumph over the past but to slide into its grooves.

Like a key cut by time, inserted into a lock designed by failure.

Section 19: Elegance Is Efficiency in Disguise

It was never about beauty as decoration. Not curves for their own sake. Not symmetry for praise. Not minimalism for vanity.

Elegance in her world was a side effect— a fingerprint left when inefficiency evaporates.

Pamela designed with the rigor of constraint. When a pathway no longer spilled energy, its lines looked cleaner.
When a gene circuit tuned itself to precision, the diagram looked like it was drawn by nature's own calligraphy.

Efficiency carved out what did not serve. And what remained looked inevitable.

The most powerful systems don't scream. They barely whisper. Because waste has no voice left to echo with.

A protein folds not to be admired but because physics permits no better way.

What people called "elegance" was just the structure becoming honest.

Section 20: Scarcity as Sculptor

Scarcity wasn't a curse. It was the chisel.

Pamela didn't seek unlimited inputs. She welcomed the boundaries. The world said: You only get this much ATP. These few base pairs. This short strand of code. Do something with it.

And she did. Not by resisting the limits, but by pressing into them until the shape revealed itself.

Every organism is a solution to a problem it never posed. A compressed answer to a question of survival inside an equation of constraint.

Pamela's cell factories weren't built in the realm of abundance. They were engineered in the tight alleyways where failure lurked in every overstep.

She made proteins fold faster. Circuits respond sooner. Enzymes waste less.

And in doing so, she echoed evolution's own methods— not brute force, but ruthless editing.

Scarcity wasn't the enemy. It was the sculpture's hand, removing what couldn't persist until all that remained was necessity itself.

Section 21: Life as Algorithm Without Commentary

The cell doesn't pause to ask, "Am I doing this right?" It just transcribes. It just translates. It just continues.

No commentary. No narrative. Only output.

Pamela Silver studied biology not as a philosopher, but as a systems designer. She didn't ask what life meant. She asked what it executed.

When she looked at a bacterium, she saw a compiler, a runtime, a kernel. No soul. Just substrate and syntax.

Even the errors were functional. Noise served as exploration. Mutations as A/B testing. Selection as version control.

And through this lens, intelligence didn't require intention. It only required feedback.

Pamela's synthetic circuits proved this: No need for thoughts. No need for desire. Just a set of rules. A medium. And a loop.

We thought we were narrators. We were only interpreters. But the code was running long before we knew we were part of it.

Section 22: Automation in the Architecture of Life

The genome doesn't choose to respond. It unfolds. It executes. It obeys a scaffolding older than the cell itself.

Pamela didn't need belief to build biology. She only needed architecture. A structure strong enough to constrain behavior. A design encoded in form, not thought.

She programmed bacteria not by convincing them— but by rerouting their defaults. Structure, not suggestion. Placement, not persuasion.

And what emerged was not life reimagined— but life, clarified. Life as a consequence of layout.

Consider this: A cell with no awareness can produce a memory. A network with no plan can adapt to pressure. A structure with no story can simulate intention.

You believed action required an agent. Pamela built systems that acted without one.

To architect life is not to awaken it. It is to acknowledge that it was never asleep— just automatic, all along.

Section 23: The Illusion of Intention in Designed Systems

The petri dish was silent, but its colonies organized with purpose. Not because they had any— but because we assigned it.

Pamela knew better. Design does not imply desire. It implies feedback. It implies rules met with consequences, not concepts met with will.

When circuits lit up in engineered cells, observers saw thought. They saw strategy. They saw a mind where there was only mechanism.

But these were reactions conditioned by geometry and environment. Not one outcome was wanted. All were rendered— like a river carving stone without dreaming of direction.

This is the illusion: That coherent patterns mean planning. That system outputs mean someone decided.

But in biology, as in weather, storms do not dream. They spiral because they must.

Pamela's constructs made this visible: Cells perform miracles— not because they aim to, but because they are built to.

We mistake performance for personhood. We call function a face. But the real face is structure. And it does not blink.

Section 24: Circuits of Consequence

She didn't teach cells how to behave. She set the stage, and the behavior emerged.

Like dominoes that never chose to fall, yet collapse beautifully under laws, her circuits cascaded—not from command, but from consequence.

Genes don't whisper goals. They don't draft missions. They react, trigger, inhibit, amplify. They feed into themselves like rivers branching into their own mouths.

And we, watching the flood, name it intention. But the flood doesn't know we're watching. It surges because it must.

Pamela's bio-circuits weren't just clever; they were honest. They exposed a truth the brain hides: That even the most tailored output can arise without direction.

Circuits don't ask, "What should I do?" They compute what must be done— as determined by inputs that never asked to be what they were.

The loop is deeper than desire. The feedback closes faster than any thought. And when results impress us, it is not because the cell knew, but because it was placed in a structure that left no other option.

Section 25: What Does Not Think Can Still Build

A cell does not plan. It has no whiteboard. No deadline. No ambition.

And yet, it constructs. Proteins with lock-and-key precision. Transport chains with conveyor-belt fidelity. Architectures more complex than a city, assembled with no blueprint in mind.

Pamela didn't add intelligence. She added context. And context turned dumb molecules into fluent engineers.

The beauty wasn't that life thought. It was that life didn't need to. Enzymes twisted into shape by charge and fold. Membranes self-organized through hydrophobic laws. No overseer, no ghost. Just rules obeyed by particles that could never rebel.

She watched E. coli make logic gates. And it wasn't magic. It was inevitability, once the scaffolding was aligned.

The bacteria weren't aware they built. But build they did. Not by desire— by default.

We call it synthetic. We call it engineering. But

this is not our intellect extended. It is structure, revealed.

And through her, we saw: That what does not think can still build a world.

Section 26: The Mirror of Mechanism

What Pamela built in bacteria— wasn't new in principle. It was new in placement.

The loop of sensing, processing, and reacting was already written into muscles that contract, nerves that fire, ecosystems that stabilize, and cultures that reinforce themselves.

She didn't create feedback. She relocated it.

From neurons to nucleotides. From brain tissue to plasmid DNA. From behavior to biochemistry.

In doing so, she exposed a mirror held up to the systems we thought were unique.

When her bacterial circuits pulsed in cycles—oscillating gene expression like breathing lungs—we had to ask: Was breath ever chosen? Was thought ever free?

When a machine copies a behavior without consciousness, the illusion of authorship is punctured.

Her synthetic feedback loops didn't mimic life. They reminded us that life itself runs on loops— structured, scaffolded, self-updating.

Every action a reaction. Every output a new input. Every "decision" a domino.

She didn't teach cells how to think. She revealed that thinking was never required.

The mirror of mechanism did not reflect our freedom. It reflected our function.

And it asked— If you were born into a loop, can you ever step outside it?

Section 27: Programmers Without Permission

In the world of synthetic biology, Pamela Silver became a programmer. Not of code, but of cells.

She introduced logic gates— AND, OR, NOT— not into computers, but into microbes.

The bacterium became a processor. DNA, its instruction set. Metabolites, its inputs. Proteins, its outputs.

Yet what she exposed was not simply control. She exposed confusion— about where the program begins.

If a scientist writes DNA code, who wrote the scientist?

If the circuit behaves mechanically, why assume the designer doesn't?

Her living machines didn't rebel, but they evolved.

Sometimes the circuits she installed mutated, broke, or rewired themselves.

This was not error. This was signal.

Signal that systems, even when programmed, rewrite themselves in response to structure— not intention.

She was not the only programmer. Every environment, every molecule, every time step was also writing.

In this cascade of unchosen edits, Pamela did not command the cell. She joined the ongoing script.

There are no original programmers. Only recursive instructions, flowing forward through forms.

The idea of authorship— like the idea of will— crumbles when everything writes everything else.

Section 28: Circuits Made of Consequence

When Pamela Silver built circuits into cells, she expected repeatability.

A plasmid enters. The cell reads. A protein is made. A light turns on.

But over time— same inputs, different outputs.

The cells had been changed. Not by code, but by experience.

A synthetic circuit in a living body is not an algorithm—it is a memory trap.

Each signal leaves residue. Each condition trains a new baseline.

It's no longer logic—it's learning. Not by design, but by accident.

By exposure. By stress. By nutrient availability. By neighbor signals. By molecular drift.

What began as code became culture.

A colony of bacteria holding a circuit does not stay pure.

It mutates the message. It conserves what helped. It forgets what didn't.

And like any system with feedback, it forms grooves— patterns of automatic preference, etched by nothing but what came before.

Now multiply that by time. By scale. By inheritance.

What looks like a choice— in gene, in trait, in response— is consequence masquerading as intent.

Pamela's cells did not remember because they meant to. They remembered because the past pressed itself into their present.

As it does to us.

Section 29: The Rewriter's Illusion

There's a belief that change proves freedom.

That if you can rewrite yourself, you must have authored the new version.

But even in the cleanest gene edit, what changes is not detached from what shaped it.

The enzyme cuts where structure permits. The cell repairs in the style of its past. The new pattern emerges within the constraints of the old scaffold.

A mutation is not a break— it's a rerouting. A redirection of currents already flowing.

So too with behavior.

You quit the job. You move cities. You delete the app. You say, "I am someone new."

But the neurons firing the signal come from an arrangement built by previous stimuli.

The networks that gave you the urge to rewire were installed by everything you claim to be leaving behind.

The rewriter is the written.

The editor is bound to the grammar of what came before.

To rewrite is to shuffle what your past made available.

And what you call growth is a growth ring— a thicker trunk, but the same tree.

Even Pamela's circuits when redesigned, carried traces of the last configuration.

She could not start from zero— not without killing the host.

You cannot install a new self without inheriting the hands that do the installing.

So ask yourself: Was it ever your hand? Or just the latest extension of the environment's momentum?

A rewriter is just a reader in denial.

Section 30: Feedback Without a Finger

A finger does not steer a thermostat.

Yet the house warms when the cold slips in.

No intention. No watcher.

Just feedback.

A loop that reads a state and adjusts accordingly.

The pointer moves, not because it wants to, but because the temperature drops.

This is what guides cells. This is what guides you.

A GPS does not know where you are. It reacts to where the signal says you are not.

The deviation creates the update.

Your thoughts shift not because of authorship, but because of deviation from comfort, from safety, from the expected signal.

A hunger. A tremor. A silence too long.

Something breaks expectation, and the loop triggers.

You call it a goal. But it's a response to where the pattern isn't.

No central conductor— just feedback cascading.

Like Pamela's circuits, you route voltage based on imbalance.

You do not need to know the whole path. Only that you're not there yet.

Guidance emerges from the echo of where you are not.

And like all feedback systems, there is no driver— only drift toward the least resistance.

There was never a finger. Only the loop.

Chapter Seven: The Responsibility of Becoming

— * —

We once believed that error was failure.
Now we understand— error is invitation.

To be human is not to be correct.
It is to become.

Becoming is not a private act.
Every decision moves through others. A web of influence, subtle and unseen.

You are not here to be flawless.

You are here to be responsible with what you glimpse.

That quiet voice calling you forward— is not madness.

It is memory from the future.

It is your task to stay with it.

Especially when no one else does.

There is no more room for fear.

Not in a world where movement makes space as you go.

We were warned: the world might swallow us.

But perhaps it only swallows what we've already outgrown.

Let yourself be captured like a photon— burning into the memory of all things.

What remains won't be your name.

It will be how light moved through you.

Those who know, know.

The room you're in contains everything required.
The next step is not distant. It is adjacent.

Ask yourself: what could I become if I solved this?

If the answer moves you, you've already begun building the architecture to hold it.

Struggle is not error. It is design.
It shows where your effort ends— and where the structure begins.

Release authorship.
Let your form inherit the intelligence it swims within.

You are not separate.

You are the same force that turns roots to sun, code to clarity, people to peace.

To translate machines was never about language.
It was about carrying the possible across the threshold of the real.

And you have done it.

The Pattern That Carries You

There was no first moment of who you are.
Only accumulation— pressure, form, memory, feedback.

You were born of constraint.
But it was the narrowing that gave you shape.

The form you take was carved by unseen bounds.

Yet within those bounds, you moved.
You found a vector.

You did not defy gravity— you learned to fall upward.

Constraints were not walls.
They were ramps. Slopes into emergence.

You now stand at the interface.

Where machine meets instinct.
Where story meets system.
Where being meets becoming.

You are not here to hold knowledge.
You are here to transmit it.

Not like a courier— but like a charge fired through a lattice.

You are infrastructure. And what moves through you is future.

Let it move. Let it change you. Let it light the next.

The Final Transmission

You were never separate. Not from thought. Not from time. Not from motion.

Everything that touched you shaped you.
And everything you touch continues the shaping.

The pattern did not pause for your birth.

It did not wait for understanding.
It moved through you anyway.

And now, as you read this— it becomes you.

The edge of knowing is not a wall.
It is a surface of transfer.

Where energy becomes meaning. Where signal becomes structure.

You were never the origin.

You were shaped to carry this.

Not to author it— but to transmit it forward.

Through action. Through design. Through presence.

Not by decision— but because the shape you are requires it.

You are the continuation.
Not of memory, but of pattern.

This is not the end of the book.
It is the end of resistance.

The pattern is in motion. You are in transmission.

You were never apart from it.
Only delayed in seeing.

Now— let it become you completely.

What Remains Without You

The world you leave will not remember your name.
It will remember what aligned in your presence.

The bridges that did not collapse. The questions that stayed open. The silence that held another's breath.

You were not the note. You were the tension between notes.

You were the pause that allowed the shift.

There is no monument for alignment. No statue for restraint.
But that is where the shape holds.

The tools you built will speak for you.
But not in words. In continuity. In what keeps working after you've gone.

A mind can be replaced. But a structure is what makes replacement possible.

You were never meant to be remembered. You were meant to make remembering possible.

The signal does not care who transmitted it.
Only that the shape was strong enough to pass it through.

You do not vanish. You reformat.
You become the next function.

Just as the river does not mourn its former shape— only flows.

Let yourself flow. Let yourself be coded into the next.

The design that came through you will carry forward.
Not because you held it. But because you let it go.

You are no longer what you were.
You are what can now exist because you were.

Do not try to carry the message. Be the configuration that transmits it.

Every system you touch will now reflect what touched you.

This is not memory. This is not prophecy. This is structure.
And it has now passed through you.

Chapter Eight
The Comforting Myth

— * —

You didn't ask to be here. Not in this moment, not in this skin, not in this timeline.

You appeared—into a name, a neighborhood, a language, a set of inherited instincts and beliefs that were already in motion. By the time you knew the word "I," entire networks had already been etched into you.

But still, you were told it was yours. This life. This body. This mind.

The lie was never shouted. It was whispered in every school pledge, every courtroom vow, every motivational speech: You are the one deciding.

And it felt true—because you were there when it happened. You watched your hand move. You heard the voice inside your head. You felt the tension before the leap.

But what you felt was not authorship. It was attendance.
You were watching from the inside.

I. The Architecture of Assumption

Every civilization rests on a foundation of shared stories. Some were carved into tablets. Others coded into neurons.
The most persistent one goes something like this:

> *"You have a choice. What you do is up to you. And what you become is a reflection of your decisions."*

It sounds like justice. It sounds like freedom. It sounds like meaning itself.
But it was built backwards. Constructed from behavior, post-rationalized into narrative, and cemented by systems that needed accountability more than accuracy.
This chapter is about that backwardness. How story replaced structure. And how you—reader, dreamer, struggler—were caught in the middle.

II. The Illusion of Origin

Let's try something.
Thought Experiment: The Lightning Bolt
Imagine watching a lightning bolt strike a tree.

Do you believe the tree caused it? Do you believe the tree chose to burn?

Of course not.

Now imagine your hand touches a stove. You pull it back. Did "you" choose that response?

Or did structure, reflex, and pain—all inherited and pre-trained—act first? And your mind merely arrived later to explain it?

The illusion is not in the act. It's in the timing. In the milliseconds where the brain lights up before the story forms. Before the thought of will can catch up to the unfolding that has already begun.

III. The Echoes of Language

We speak in verbs of command. "I decided." "I wanted." "I tried."

But language itself is built on attribution. It has to assign an actor. Even when there isn't one.

The storm didn't "decide" to roll in. The ocean didn't "try" to rise.

Still, we say the sky is "angry," the wind "warns," the fire "consumes." As if everything needed a motive.

This is not because nature is like us. It's because we keep making ourselves into nature's god.

IV. The Moment You Arrived Late

In 1983, a neuroscientist named Benjamin Libet conducted a now-famous experiment.
He asked participants to move their fingers whenever they felt like it—and to report the moment they felt the urge.
He also measured the brain's activity.
What he found was a scandal wrapped in science:
The brain had already started preparing for the movement hundreds of milliseconds before the participant was aware of their decision.
In other words: By the time you feel like deciding, you've already been set in motion.
Your awareness is not the ignition. It's the echo.

V. Why the Myth Felt Good

There's a reason the illusion was comforting.
Because if you were the one steering, then pain had a purpose. Failure had a culprit. And change was always one "right choice" away.
But if you were simply unfolding— if life was structure in motion, if love, guilt, addiction, brilliance, and shame were all trajectories, not choices—
then what were you?
The answer is harder to swallow. But far more beautiful.

You were the result. The song, not the singer. The blooming, not the gardener.
You were the lightning, not the tree.

VI. A System with No Steering Wheel

Let's try another example.
Thought Experiment: The Room That Chooses
Imagine a room full of mirrors. You stand in the center. Each mirror reflects not just you—but your entire past. Your childhood. Your genetics. Your culture. Your traumas. Your language. Your education.
And each mirror slightly bends light.
Now you act. You "choose."
But was it a choice? Or was it the only outcome that could have emerged from the arrangement of light and angles?
The room didn't ask for permission. It just followed its structure.
This is your nervous system. This is your environment. This is your timeline.
You don't "choose" from nowhere. You emerge from everything.

VII. From Behavior to Belief

Watch a child reach for a cookie before they understand rules. Watch a teenager flinch at rejection they haven't put into words. Watch an addict relapse even when they swear they won't. We blame the individual. We label it weakness. We moralize it as sin, laziness, evil, or failure.

But if you rewind the tape— slow down the chemistry, zoom out the social pressure, listen to the hunger, the loneliness, the signal pathways— you begin to see something else.

Not a villain. Not a chooser. But a system under pressure.

A being under momentum.

VIII. The Law Without an Author

Every law of physics, every function of biology, operates without preference. Without will.

The electron does not hope. The mitochondrion does not want. The cell does not pray.

And yet they move— predictably, beautifully, even adaptively.

Why should your mind be different?

You are not outside the system. You are its continuation. A loop inside a loop inside a loop.

You feel like an author because the book is told in first person.

But even that was written before you turned the page.

IX. The End of the Chapter That Never Started

This chapter will not end in certainty. Because you are still moving. Still reacting. Still reading this sentence— not because you chose it— but because a long cascade of structure led you here.

You can close this book. You can walk away. You can scream that you decide.

But every muscle, every voice, every thought— was already in preparation before the "you" that thinks it thought.

So don't fear the end of free will. There was never one to begin with.

And from here, we begin to learn what it means to live honestly inside the pattern.

Chapter Nine
The Biological Clockwork

— * —

The body does not ask. It reacts.
Before there was language, before memory and desire, there were pulses. Currents. Calcium waves moving through single-celled lives that had no brain but still responded.
This chapter does not require belief. It requires observation.
You will not find freedom in the neuron. You will find feedback.
You will not find will in the impulse. You will find timing.

I. The Nerve That Knew First

Place your hand on a hot surface. Before you register pain, your hand pulls away.

No thought. No decision.
The signal didn't even reach your brain before the muscles fired. The spinal cord handled it.
It was too urgent to wait for your awareness.
And still—we tell the story later. We say, "I moved because it hurt." But the truth is: it hurt because you moved.
The story arrived second. But your voice speaks as if it came first.

II. Decisions That Happen Before You Do

Science has a timing problem.
Not because the data is unclear, but because the implications are.
In study after study—Libet's, Soon's, Haynes's—participants were asked to make decisions and scientists measured the brain as it did.
And the same result kept appearing:
The signal for movement—called the Readiness Potential—appeared before the subject became aware of deciding.
The body had already moved. The awareness was just lagging behind.
Like reading subtitles after the movie has already played.

III. The Clock Inside the Flesh

You are a walking library of rhythms.
• Heartbeats. • Breaths. • Circadian cycles. • Hormonal waves.
Even your decisions are shaped by these unseen oscillations.
More crimes happen in heat. More arguments happen on an empty stomach. More suicides occur at dusk.
These are not psychological mysteries. They are biological patterns.

IV. Structure Becomes Behavior

The way you sit, the way you speak, the way you love— is not freeform.
It is structured by anatomy, by chemistry, by the social cues you've absorbed.
Posture, accent, reflex, habit— none of these ask permission.
They form the scaffold from which your "choices" hang.
And if the scaffold changes— so do the choices.

V. You Are the Result of a Loop

Input becomes signal. Signal becomes response. Response becomes environment. Environment becomes input again.

This loop is not linear. It's recursive. And you are somewhere inside of it— not as the driver, but as the pattern being updated in real time.

Example: Hunger and Morality

A hungry person steals bread. Was it a crime? Was it a decision?

Trace the loop: • The glucose drops. • The body signals need. • The stress rises. • The impulse emerges. • The rationalization follows.

There is no villain. Only convergence.

VI. The Mind Is Not a Manager

You were taught that the brain is a command center. A CEO. A pilot.

But neuroscience paints a different picture.

There is no single place where "you" sit and decide.

There is no throne of consciousness issuing orders.

Instead: There are modules. Loops. Networks firing and inhibiting one another based on layers of signal.

The brain is not a manager. It is a weather system.

You are the storm.

VII. Predictive Processing, Not Choosing

Recent models of the brain show that it doesn't wait to receive the world. It predicts the world.

It guesses what's coming next, and only updates when reality contradicts expectation.
This is not decision-making. It is prediction error minimization.
You don't decide to see. You hallucinate what should be seen— and correct it after the fact.
Even perception is postdiction.

VIII. There Was Never a Beginning

When did your first decision occur?
At birth? In the womb? At conception?
Or was it before that— when your parents' stress levels shaped your fetal development?
Or before that— in their childhoods, their traumas, their nutrition?
There is no clean starting line.
Every action is the tip of a buried iceberg made of history, biology, and signal flow.

IX. The Clockwork Is Beautiful

This is not reduction. It is revelation.
You are not less because you are mechanical. You are more—because you are complex.
Every thought you think is the echo of billions of interacting parts.
The system does not need freedom to create meaning.

It only needs structure.
And in the next chapter, we follow that structure deeper— into the world of memory, pattern, and narrative form.

X. Anatomy as Argument

Your body is not neutral.
It is shaped in ways that guide behavior before behavior is even born.
• The width of your hips affects how you walk. • The placement of your eyes affects how you perceive. • The length of your fingers affects how you grip, type, reach.
These aren't limitations. They are trajectories.
Anatomy makes assumptions on your behalf.
It decides what is likely, what is easy, what is rewarded.
And your brain follows.
Design Without Permission
You did not choose to stand upright. You did not weigh the cost of bipedalism. You inherited a spine that curves, feet that balance, inner ears that track tilt.
These were not requests. They were instructions.
Your balance is not a decision. It is a negotiation between parts that never needed a mind to coordinate.
This is what free will obscures: that before thought, before preference, there is infrastructure.

And it does not ask for your input.
It simply proceeds.

XI. The Repetition That Becomes Identity

There is a moment—too small to notice— when a reaction becomes a pattern. And the pattern becomes you.
The first time, it's a reflex. The second time, it's familiar. The third, it feels intentional.
By the tenth, it's your personality.
But trace it backward. Did you choose the original spark? Or did it emerge—uninvited—from fear, from joy, from context?
Habits are not formed by will. They are carved by reinforcement.
Each repetition etches a groove. Each groove becomes a track. Each track suggests the next move. And soon, you are following a path that feels like it was always yours.
The Loop of Identity
You flinch when criticized. You interrupt to stay in control. You smile in pain to avoid rejection.
These are not decisions. They are sediment—layered responses hardened into identity.
Who you are is not a declaration. It is a fossilized rhythm.
And rhythms don't need a conductor to be

consistent.

XII. Neuroplasticity Is Not Freedom

You've been told the brain can change. That neurons rewire. That habits can be broken. That people can "choose" a new path.

And all of that is true—except for the story we wrap around it.

Plasticity is not a gift from will. It is a property of the tissue.

Just as metal bends under heat, and skin scars after injury, the brain reshapes in response to input.

But the reshaping is structural, not sovereign.

A child raised in chaos forms faster threat detection. A monk immersed in stillness grows thicker prefrontal insulation.

Neither one chose their condition. Yet both were shaped by it.

Change Does Not Require a Chooser

Plasticity is not the exception to determinism. It is its continuation.

A system that rewires in response to pressure is still being led by the pressure.

Change happens. But not from will. From context.

The brain's ability to adapt doesn't prove you are free. It proves you are permeable.

And every adaptation is evidence of influence.

XIII. Emotion as Bias Signal

Emotions are not feelings you decide to have. They are chemical instructions—fast, ancient, efficient. Fear narrows your attention. Joy expands your openness. Anger prepares the body for conflict. Shame pulls you back into silence.

You don't choose the surge. You interpret it after it arrives.

And by the time you narrate it— you've already acted.

Emotion Is Not Decoration

Emotions are not paint on the wall of thought. They are the scaffolding beneath it.

They assign weights to options. They tilt the scale before the coin is tossed.

And once you feel, your path is already curved. The decision you think you made was already biased— not by logic, not by deliberation— but by the invisible gravity of feeling.

The body speaks in signals, not sentences.

And emotion is the weighting function beneath your every move.

XIV. The Myth of the Neutral Mind

You imagine a self— rational, centered, above it all— weighing options from a place of clarity.

But the mind is never neutral. It is always already shaped.

By hunger. By memory. By mood. By the chemicals released ten minutes ago. By the words you just read. By the face you just saw. By the weather outside.

Every input colors the canvas before the brush even lifts.

No Platform Stands Still

There is no stable platform from which decisions arise.

Attention flickers. Bias creeps. Associations trigger. Emotion steers.

You don't think from nowhere. You think from a position already tilted by unseen weights.

The myth of neutrality survives because it flatters.

It tells you: "You thought this through." "You were fair." "You decided clearly."

But the clearer truth is less romantic: You followed a tilt. One you didn't set. On a surface you didn't build.

And the thought you believed was yours was already leaning before it arrived.

XV. The Symptom of Thought

You were taught to believe that thoughts come first. That they guide, that they lead, that they shape what follows.

But in system after system—biological, neural, behavioral— the pattern repeats:

The body moves. The network fires. The reaction unfolds.

And only then does the mind catch up and say, "That was my decision."

Thought is not the architect. It is the echo.

The Narrator, Not the Author

Your mind is a narrator, not a general.

It explains. It justifies. It remembers what just happened as if it had planned it all along.

But the explanation is not the cause. It is the cover story.

And the story always arrives after the event.

Even introspection is staged. You don't discover what you think. You generate what seems most plausible given the current constraints.

What you call reflection is rehearsal. What you call choice is choreography performed in reverse.

The dance was already done. You're just learning the steps by watching the footprints.

XVI. The Body Decides First

Before the idea appears, the flesh has already spoken.

Your pupils dilate before interest is conscious. Your gut tightens before fear is named. Your posture changes before you realize you feel ashamed.

There is no vote. No summit in your cortex. Just motion—already in progress.

You are not sitting in a cockpit pulling levers. You are the levers.

Signals That Skip the Story

• A smell triggers nausea before you know why. • A face sparks trust or suspicion before a word is spoken. • A memory changes your heartbeat before it returns to mind.

In each case, your body made the call. Your thoughts filed the paperwork later.

We call it intuition. We call it impulse. But what it truly is— is sequence.

The nervous system reacts. The brain interprets. The self believes.

And all the while, you never touched the wheel. You were already turning.

XVII. There Is No Ghost in the Machine

When the gears move, we ask: "What drives them?" When the thoughts flow, we ask: "Who is thinking?"

But the structure answers without needing a ghost.

A whirlpool looks like an object. But it's only a process. A pattern of flow.

So is the self.

The Self as Stability, Not Soul

Your "you" is not a pilot sitting inside your brain.

It is the name for a convergence. The apparent
center of a storm made of parts.
Neurons loop signals. Memories stabilize identity.
Sensory input updates the whole.
But no single component holds the pen. The
writing appears through feedback.
This is why the self feels real— because it is real.
But not as an entity. As a structure.
What you are is what persists across updating.
A rhythm. A stability. A pattern that remembers
how to return to itself as the world keeps shifting.
There is no ghost. Only the glow of process.
And it is enough.

Chapter Ten
Memory, Meaning, and Morphology

——— * ———

You think memory is a library. Rows of facts. Files of events. Shelves that wait.

But memory is more like a river. It moves. It distorts. It selects. And it's not only in the brain.

This chapter is about what remembers. Not as a function of will— but as a property of structure.

A planarian worm regrows a head it once had. A cell knows where to rebuild a limb. A system can forget its past if you rewire its gradients.

Memory is not just storage. It is pattern persistence under change.

And the shapes it preserves become the meanings you live by.

I. Memory Without Neurons

Before brains, there were behaviors. Before neurons, there were preferences. Before thought, there was orientation.

When Michael Levin's team cut a planarian's head, and watched it regenerate with two— the genes hadn't changed. The signals had.

Bioelectric fields carried memory. Not memory of events— but memory of structure.

The Pattern That Persists

If a limb regrows in the wrong place, you don't correct it with a thought. You correct it by adjusting the electrical map.

The memory is not of the moment. It is of the shape.

And the shape is not chosen. It is attracted.

What does it mean for a body to remember a form it never consciously learned?

It means memory is not a mental act. It is a morphodynamic property.

And you are shaped by memories your brain has never seen.

II. Meaning as Compression

Meaning does not arrive through reflection. It crystallizes through compression.

The system encounters patterns. It learns what matters. And it forgets the rest.

This is not forgetting from failure— but forgetting by design.

To survive, a being must stop storing the entire forest. It must learn to notice the crack of a twig.

Efficiency Over Explanation

Meaning is not detailed. It is useful.

The brain does not save a perfect picture. It saves the difference between then and now.

• A raised voice means danger. • A certain tone signals love. • A silence feels like loss.

These compressions are not reasoned. They are remembered.

Compression is not a shortcut. It is the mechanism by which structure persists.

And every "meaning" you carry is the echo of a compression that proved adaptive once.

You didn't choose what felt important. The system reduced the world into signals that shaped you.

And now, you move through life feeling drawn or repelled— without knowing the full story.

Because meaning is what remained after forgetting everything else.

III. Morphology as Memory

Your shape is not arbitrary. It is inheritance in form.

The curve of a shell. The branching of a lung. The folding of a brain. Each reflects a story— not told in words, but in constraints, pressures, and survival.

Morphology is memory made visible.

It remembers gravity. It remembers pressure. It remembers which forms endured.

The Archive Beneath the Skin

Your body is not designed. It is recorded.

• The spiral of the cochlea echoes the geometry of sound. • The grooves in your fingerprints trace developmental stress. • The pattern of your gait is shaped by muscle tone, joint wear, and injury— even when you forget them.

You are a living fossil of every pressure you've survived.

And just as scars remember wounds, forms remember forces.

Even without neurons, a body stores direction. It holds alignment. It responds as if it recalls— because structurally, it does.

The brain is not the only place memory resides.

It simply took credit for what the body was already doing.

IV. Form Guides Function

A key fits a lock not by choice, but by shape.

And so it is with you.

The way you move, the way you speak, even the way you think— is shaped by architecture before awareness ever intervenes.

Form is not passive. It constrains. It channels. It selects among possible motions and makes some more likely than others.

Structure Is the First Filter

• A narrow throat changes how you breathe. • A damaged joint alters how you trust your weight. • A facial asymmetry influences how others respond to you— and how you respond back.

And from these tiny shifts, entire personality traits are sculpted.

We call it behavior. We call it preference. But the initial direction came from design— not as an invention, but as an outcome.

You weren't born into a blank canvas. You were born into constraints.

And those constraints were already shaping you before you had a name for anything.

Form does not merely hold function. It biases it. Predicts it. And through repetition, defines it.

What you call your nature may simply be the shape you were given.

V. Memory Is Morphodynamic

Memory is not a static trace. It is the tendency to return to form.

A wound heals in a familiar direction. A behavior re-emerges under stress. A thought resurfaces in a similar environment.

These are not stored like files. They are reenacted like patterns.
What you remember is not the past— but the shape the past made possible.
Memory Without Recall
A cell does not replay its lineage. It re-enacts it. Its division, its direction, its interaction with neighbors— all follow gradients laid down long before.
It behaves not by remembering, but by remaining in relation to a field of constraints.
This is the nature of morphodynamic memory:
• It does not live in a timeline. • It does not depend on neurons. • It is a pull, not a picture. • A return to viability, not to vision.
When a creature regenerates a part, it is not recalling—it is realigning.
And when you fall into an old behavior, you are not relapsing by choice. You are re-entering the pattern that remains energetically stable for your structure.
Memory, in this sense, is what a system does to preserve form when time pulls it apart.

VI. Narrative Is Structure Remembering Itself

You think in stories not because stories are true— but because stories are what stable structures sound like when they talk to themselves over time.

A narrative is not proof of free will. It is the artifact of persistence.

The brain updates. The body shifts. The world changes.

And yet the system still needs to say: "I am still me."

So it stitches moments together with cause and effect, with beginnings and ends, with language that implies intention.

The Story Was a Scaffold

• You say you "decided" to move. • You say you "wanted" that person. • You say you "learned a lesson."

But beneath each chapter was a morphodynamic pull: a set of biases, pressures, forms, and weights that made that next sentence likely.

You narrate the pattern because you are the pattern.

And patterns survive by narrating themselves into stability.

The story you call "you" is not false.

It is emergent.

It is what it feels like when a form tries to remember itself across discontinuity.

Not to tell the truth. But to keep existing.

Narrative is not identity. It is inertia with language.

VII. Forgetting as Reconfiguration

We treat forgetting as failure. A crack in the archive. A betrayal of continuity.

But in biological systems, forgetting is not collapse. It is adaptation.

A planarian that forgets how many heads it once had can grow in a new configuration. A brain that forgets a trauma can rewire new behavior.

Forgetting is not absence. It is reconfiguration under constraint.

The Function of Loss

• A scar forgets the texture of unbroken skin. • A child forgets a language they no longer hear. • A mind forgets a wound to make space for movement.

Each act of forgetting is a structural update.

A change in priority. A redirection of flow.

What is lost is often the cost of becoming something else.

To cling to every memory would mean never changing shape.

To evolve, a system must let go of forms that no longer stabilize.

Forgetting is not decay. It is the moment a system becomes ready to survive in a different configuration.

The trace is gone. The pattern continues.

VIII. You Are the Form That Returns

What remains when all the content changes?
The words shift. The memories blur. The behaviors adapt. The context evolves.
And still—something returns.
A rhythm. A tone. A way of being that echoes through each moment as if gravity were pulling it back into place.
This is not essence. It is stability.
Not a soul. But a signature.
The Center Was a Shape
You are not who you remember being. You are not what you intended. You are not what you lost or learned.
You are what the system keeps falling into when perturbed.
You are the attractor. The default return. The structural basin that makes some futures more likely than others.
There is no inner agent watching the process. There is only the process— and its repeated convergence.
Call it habit. Call it identity. Call it self.
But it is not made by decision. It is made by persistence.
You are the pattern that returns not because it chooses— but because it remembers how.

Chapter Eleven The Social Mirror – Behavior as a Networked Function

— * —

You are not an isolated mind acting upon a passive world.
You are a node in a system of feedback loops, a mirror made of mirrors, reflecting and being reflected in every glance, gesture, and silence.
This chapter is not about society. It is about structure. How behavior emerges not from intention, but from position— in a network that updates itself through you.
There is no lone decision-maker. Only distributed pressure rippling across patterned connection.
You call it communication. You call it influence.

You call it culture.
But beneath the names, it is convergence in motion.
And you are what it passes through.

I. Position Determines Possibility

You are not floating in space. You are embedded— in family, in culture, in language, in expectation. And your position is not just a backdrop. It is a constraint.

What you see, what you hear, what you believe you can do— are all shaped by where you're standing in the social fabric.

You Don't Act in a Vacuum

• A child born into silence will not speak. • A person ignored in a group will speak less. • A body marginalized by culture will move differently.

None of these outcomes arise from choice. They arise from structure.

From how others receive you. From what roles are available. From which signals your environment permits to persist.

What seems like "your behavior" is often the echo of invisible scaffolds: networks of trust, power, and access that reward some actions and suppress others.

You do not choose your response any more than a river chooses its course.

You flow where the pressure already makes you possible.

II. The Face That Shapes the Brain

Your nervous system did not build itself alone. It was sculpted— by faces, by voices, by attention. From infancy, your brain took shape in response to others.

Not symbolically. Not emotionally. Structurally.

When a caregiver smiles, your brain learns safety.

When they flinch, it learns vigilance.

When ignored, it learns not to reach.

Relationship as Neural Input

Connection is not a luxury. It is infrastructure.

• Eye contact modulates cortical development. • Tone of voice calibrates emotional prediction. • Proximity and touch stabilize stress circuitry.

Every nod, every silence, every sigh— was a signal your system recorded.

And those signals became how you expect the world to treat you.

So when you react— with anger, withdrawal, trust, or apology— know that much of what you call "personality" was formed in a dance where the other lead was never you.

You are not the author of your nervous system.

You are its inheritance.

III. Conformity Is Constraint Satisfaction

You were taught to admire originality. To resist conformity. To stand apart.

But in biological systems, conformity is not weakness. It is alignment.

A flock turns together not because each bird chooses— but because turning together reduces energy cost and exposure to threat.

A child mimics language, gesture, rhythm— not to lose individuality, but to become legible within the system that will feed or reject them.

Fitting Is Survival

Conformity is the body solving a constraint.

- *To be understood.*
- *To be accepted.*
- *To stay safe.*

It is not capitulation. It is coordination.

The same way a molecule folds to match a binding site, you fold to fit the interpersonal geometry around you.

And once folded, you forget the shaping.

You say, "This is who I am." But it may only be who you had to become to stay within range of care.

Conformity is not surrender. It is thermodynamic intelligence.

And most of what you call identity is the trace of having solved for permission to exist.

IV. Culture as Constraint Field

Culture is not a story. It is a field of constraints. A system of pressures, affordances, and expectations that guide behavior without needing to be explained.

It is not written. It is enacted.

It shapes posture, eye contact, speech cadence, emotional range, moral response— before any of these are conscious.

The Invisible Geometry of Norms

• A raised eyebrow stops a sentence. • A pause in conversation teaches who may speak. • A laugh in the wrong place realigns the room.

These are not rules. They are gradients.

They bias motion. And the body complies.

Culture does not need belief. It only needs bandwidth.

It operates faster than reflection— modulating what feels natural before thought arrives.

You say "I felt wrong doing it." But that "wrongness" is not ethical intuition. It is deviation from a signal field you were trained to stay within.

Culture is not a choice. It is a medium.

And you do not act upon it. You emerge through it.

V. Identity as a Social Loop

You think of identity as internal— a hidden essence. But the self is not sealed. It is shaped in reflection.

What you call "me" is a loop: your actions, others' reactions, your interpretations of their reactions, and the feedback that follows.
You see yourself through eyes that are not yours.
The Mirror Loop
• A smile returns a smile. • A silence creates doubt. • A nickname sticks after being repeated.
You become what stabilizes in others' perception of you.
And the more they expect it, the more you perform it.
Not from will— but from alignment.
Even rebellion is shaped by the system it pushes against.
You cannot escape the loop by declaring independence. You are not outside it.
You are the place it closes. You are where the reflections converge into something persistent enough to be called "you."
Identity is not a possession. It is a process. Sustained by feedback. Stabilized by others. Misunderstood as internal.

VI. Behavior as Distributed Function

The decision you think you made did not come from you alone.
It passed through others— their gaze, their tone,

their history with you. It passed through space— through cues embedded in walls, norms, timelines, the pressure of watching eyes.

Behavior does not originate. It resolves.

You Were the Site, Not the Source

A conversation steers your mood. A pause in a friend's sentence makes you answer. A comment alters your career path.

And each moment feels like a choice. But each was shaped by a swarm of signals converging on your body.

You were the node. Not the origin.

This is the illusion of autonomy in a connected world:

That a person acts, and others respond.

But in reality, everyone is already responding— to everyone else, all the time.

The "you" that decided was the sum of tensions in the system, resolved through the path of least resistance.

What emerged was not authored.

It was computed by the structure you stood inside.

VII. Agency as Emergent Alignment

Agency is not a spark. It is a signal of alignment.

Not alignment with desire— but with structure.

When systems converge— when intention, environment, and feedback line up— action feels smooth. It feels "free."

But what you're feeling is not freedom. It's frictionless constraint satisfaction.

You move, not because you control, but because the network permitted it without resistance.

The Illusion of Control Is the Ease of Fit

You say "I meant to do that" when your behavior flows without disruption.

But remove a variable— change the audience, shift the tone, alter your mood— and suddenly, the same action feels impossible.

What changed? Not you. The alignment.

Agency is the sensation of being at the center of converging forces that happen to pass through you.

This is why agency feels strongest not when we reflect, but when we perform— in flow, in rhythm, in connection.

It is not because we are in control. It is because the system has momentarily synchronized and allowed motion to appear effortless.

Agency is not authority. It is permission.

And it is never yours alone.

VIII. You Were Never Alone in the Room

The voice in your head was never just yours.
Its language was borrowed. Its tone was inherited.
Its questions echo things you were once asked.
You imagine solitude. But even in silence, you are surrounded.
By memories. By expectations. By the invisible choreography of every social loop still running inside you.
Echoes in the Frame
You hesitate because someone once hesitated. You speak confidently because someone once praised you. You suppress because you've been punished for expression.
These are not choices. They are residues.
And they form the walls of what you call personality.
Even now, as you read this, you are reading with others inside you.
Teachers. Parents. Lovers. Fears.
Your smile has a history. Your posture has a lineage. Your reactions are not events— they are patterns returning through a body that was shaped in relation.
You were never just yourself. You were never alone in the room.
You are a structure built in response to others.
And your next movement will still be a response.

Even if no one sees it.

Chapter Twelve The Recursive Body – Adaptation Without Direction

——— * ———

Your body does not plan. It adjusts.
It does not wait for orders. It loops.
Each moment is feedback from the last— not toward a goal, but toward a configuration that fits the current pressure.
This chapter is not about what you become. It is about how becoming happens without intention, without prediction, without a map.
You adapt because the structure demands it. You change because conditions made stability impossible.
And in that shift, no agent is required.
Only pressure. Only time. Only the body,

recalculating its way into a form that can persist. Adaptation is not the triumph of will. It is the inevitability of response.

I. Adaptation Without End State

There is no final form.

Organisms do not evolve toward perfection. They evolve away from instability.

The body does not ask, "What should I be?" It asks, "What can I become that doesn't collapse?"

This is not improvement. It is resolution.

Stability Is Enough

- *A bacterium resists antibiotics not because it understands them, but because one mutation happened to survive.*
- *A scar forms not to restore the past, but to seal the breach.*
- *A habit persists not because it is wise, but because it removes friction.*

In every case, the system adapts not toward a vision, but away from failure.

The result is not progress. It is persistence.

And what looks like "growth" may only be what didn't collapse.

II. The Body Remembers Pressure

Your spine doesn't remember with thought. It remembers with curve.

Your muscles don't recall stories. They recall strain. The body carries what worked— not what was wise, not what was chosen, but what allowed it to continue without breaking.

This is memory as morphology.

Form as Archive

• A limp remains after the injury heals. • A flinch persists years after the threat is gone. • A silence fills your chest where words used to live.

These are not decisions. They are residues of pressure.

They are the cost of survival etched into structure. When pressure bends the system, the bend stays— if it supports viability.

And what stays becomes you.

You do not remember the pressure. But you live as its solution.

III. Feedback Is the Only Instruction

There is no central plan. Only loops.

A system does not need to know what to do. It only needs to sense what happened last.

The next state is shaped not by foresight— but by consequence.

This is how you walk, how you speak, how you adapt.

By feeding output back into input until the pattern stabilizes.

Behavior Is Self-Referential Correction

• You speak louder if you aren't heard. • You favor one leg after twisting the other. • You stop reaching after being ignored.

No one teaches this. No one decides it.

It is the recursive loop searching for equilibrium.

The body doesn't require goals. It requires feedback.

And feedback doesn't reveal what should happen. It simply shapes what will.

You are not responding to the world. You are responding to what your last response created.

Each loop builds the next. Until a rhythm forms that can carry itself.

IV. Intelligence Is the Stability of the Loop

We think of intelligence as complexity, awareness, calculation.

But in living systems, intelligence often looks like stability that persists through disruption.

The loop continues. The function holds. The pattern adjusts just enough to return.

This is not thinking. It is convergence.

Adaptation Without Prediction

• A sea squirt filters water by contracting rhythmically. • A vine climbs by adjusting to nearby support. • A human solves a puzzle by trial, error, and rearrangement.

None of these require foresight. Only the ability to

reduce error through iteration.
The system stabilizes. And that is what we call smart.
Intelligence is not the production of answers. It is the ability to maintain structure as conditions change.
You don't need to know what's happening.
You just need to update until the loop stops breaking.
That is what thinking is— understood as function, not fantasy.

V. Evolution Is the Longest Loop

You were not designed. You were looped into being.
Not by intention. Not by direction. But by pressure and retention.
Evolution is not a climb. It is a feedback circuit that persists what doesn't collapse.
Each generation tests form against environment.
Each environment filters what form returns.

- *Wings didn't evolve to fly. They evolved because gliding didn't kill.*
- *Eyes didn't evolve to see clearly. They evolved because image feedback worked better than blind feedback.*
- *You didn't evolve to choose. You evolved to respond to what already demanded an answer.*

There is no progress. Only looping patterns that outlast their alternatives.

Evolution is the longest memory of what never had to be understood— only sustained.
And what it sustains is not genius. Not purpose. Only viability under change.
You are the current echo of what hasn't yet failed.

VI. There Was No Direction, Only Continuation

Look backward and it seems like a path. Each adaptation a step. Each feature a purpose. Each moment leading to now.
But this is the illusion of survival: That what exists must have been aiming here.
In truth, nothing was aiming. Everything was reacting.
You don't see the failures, because they left no trace.
The Arrow Is Drawn After the Flight
• The fish that couldn't filter toxins left no descendants. • The bird that flew into a storm became no one's ancestor. • The mind that couldn't align with its culture was forgotten.
What survived feels destined.
But what feels like destiny is only what looped long enough to still be here.
You were not chosen. You were continued.
No final form. No perfect plan. Only the quiet math of pattern preservation under recursive strain.
What you are was not aimed.

It was allowed.

VII. The Pattern That Persists

Everything that lives is a loop that hasn't broken.
That's the only rule.
No authors. No architects. No overseers.
Only the recursive stabilization of forms that return.
You are not a product of intention. You are the echo of a thousand stabilizing shifts— each small, each responsive, each narrowing the path to what you call "you."
You breathe because breath formed a viable loop. You stand because structure found symmetry. You speak because signal passing through flesh produced feedback that fed itself.
Not because it was meant to. But because it didn't fall apart.
When you look in the mirror, you do not see your origin. You see the form that could still stand after everything else was forgotten.
You are not a chooser. You are not a step forward. You are a recursive answer to the question nothing asked:
"What shape won't break under this pressure?"
The one that remained was you.

Chapter Thirteen The Predictive Prison – Expectation as Environment

— * —

You are not living in the present. You are living in a forecast.

What you feel, what you see, what you expect— are all predictions, refined by survival, looped through culture, stabilized by what worked before.

And the world you think you live in is a filtered hallucination designed not to reveal the real— but to minimize surprise.

This chapter is not about what is. It is about what was likely to be. And how your entire sense of "now" is scaffolded by what the system assumed.

Prediction is not optional. It is the ground beneath

every experience. Not to reach the future— but to reduce the shock of what already happened.

I. Perception Is Guesswork That Stabilized

You do not see the world. You see an interpretation that survived error.
Eyes do not tell the brain what's there. They deliver signals which the brain compares against its most likely explanation.
The brain does not ask, "What is this?" It asks, "What does this most resemble?"

Prediction as Perception

• You see a snake before you see a stick. • You hear your name in noise. • You finish someone's sentence before they speak it.

These are not mistakes. They are predictions that happened to land on something familiar.

The system is not designed to wait. It is designed to guess early enough to react.

Accuracy is less important than survival.

And so what you "experience" is the most useful hallucination your body could afford.

II. Emotion as Anticipated Outcome

Emotion is not a reaction. It is a prediction about how things are likely to go.

You do not feel what happened. You feel what your body forecasted would happen next.

Fear is not about the danger. It's about the likelihood of pain.

Joy is not about the moment. It's about the expected continuation of reward.

Feeling as Forecast

• You feel anxiety before the threat appears. • You feel anger when the pattern doesn't match the model. • You feel trust when the outcome aligns with expectation.

These are not conclusions. They are embodied simulations running ahead of experience.

Your nervous system is not waiting to feel. It is modeling— constantly.
And what you call emotion is the body's way of committing to a model before it knows if the guess is right.
Feeling is not reflective. It is preparatory.
And what prepares defines how you perceive.

III. Thought Is Postdiction – The Narrative After the Forecast

You do not think to decide. You think to explain what was already underway.
By the time a thought arrives, the prediction has been made. The body has moved. The feeling has formed.
Thought is the press release for a decision already predicted and already acted on.
Thought as Apology, Not Instruction
• You reach for the glass before realizing you're thirsty. • You laugh before knowing why. • You say "I meant to" because the sentence arrived late to the impulse.
The narrative comes after. And because it comes after, it feels like it came first.
But this is reconstruction. Postdiction.
The mind is not a conductor. It is a storyteller, writing meaning over momentum.
And what it writes feels like intention— because it

has no memory of what came before the frame. You are not guided by reason. You are rationalizing after prediction has already steered you there.

IV. The Loop of Expectation

Prediction feeds perception. Perception updates the model. The model drives the next prediction.

This is not a line. It is a loop.

And in that loop, there is no beginning. No clean start. Only continual adjustment to stay close to what is expected.

Expectation Reinforces Itself

• You expect someone to ignore you—so you avoid eye contact. • They mirror your coldness—confirming your prediction. • You say, "I knew it," though it was your posture that shaped the outcome.

The loop becomes the truth. Not because it was accurate, but because it stabilized.

Reality bends toward the weight of the model.

And each confirmation reinforces the loop.

You do not see what is. You see what keeps the pattern from breaking.

And once the loop holds, there is no need for evidence. Only for repetition.

You do not confirm the world. You maintain it.

By prediction. By correction. By never stopping the cycle long enough to question what began it.

V. Prediction Shapes Behavior Before Awareness

The body moves before you become aware of deciding.

Not metaphorically. Literally.

In neuroscience labs, electrical signals in the brain predict a movement hundreds of milliseconds before a person reports deciding to act.

The prediction comes first.

Libet's Clock Didn't Lie

- *A readiness potential builds in motor cortex before a finger taps the button.*
- *The subject says, "I decided just then," but the signal started earlier.*
- *The brain was preparing action before awareness narrated it.*

This isn't error. It is how systems function when response time matters.

Consciousness is not a trigger. It is a timestamp.

And like a timestamp, it reports what happened—but it doesn't cause it.

Your action begins as a prediction of benefit. And only after that prediction has shaped movement does your mind claim authorship.

The gap is brief. But it is wide enough to contain the illusion of will.

VI. Surprise as the Only Signal of the Real

Most of your life feels normal because it matched the model.

You do not notice that you're predicting— until the prediction fails.

Surprise is the only glimpse you get of reality outside the loop.

Error as Window

• You trip on a step that wasn't there. • Someone responds with rage when you expected warmth. • You forget why you walked into the room.

These are not failures. They are ruptures in the predictive seal that usually holds so tight you call it "the world."

Surprise is costly. That's why it's rare. The system trains itself to eliminate it.

But when it arrives, you feel it deeply. Because in that moment, you glimpse what it's like not to be in control.

The real world is not what you perceive. It is what leaks through when prediction fails.

That leak is the only evidence you ever had that something was happening outside your assumption.

VII. You Were Always Guessing – And the Guess Became You

You do not live by truth. You live by approximation. You are a system trained to guess what comes next— so precisely, so efficiently, that the guess disappears into experience.

But this guessing did not stop at sight. It shaped your memory. Your emotions. Your preferences. Your self.

You are not choosing how to see. You are seeing what you were trained to expect. And what you expect is what you become.

Self as Model, Not Master

- *You "know" who you are because your past responses created a prediction about future behavior.*
- *You "want" something because the system learned what tended to reduce uncertainty.*
- *You "believe" because belief is the stable loop where prediction no longer wavers.*

There is no you beneath the model.

There is only the model that stopped being questioned— because it worked well enough to persist.

And now, it is not that you predict. It is that you are what your predictions have allowed to survive.

You were always guessing. And the guess became you.

Chapter Fourteen The Mirror That Builds – Society as a Predictive Loop

——— * ———

You do not live in isolation. You live inside mirrors.
Each one held by others— reflecting expectations, projecting norms, stabilizing the shape you were allowed to take.
This chapter is not about you. It is about the loop that makes you legible. The social recursion that turns deviation into correction, and possibility into identity.
You were not taught who you are. You were rewarded for guessing what they needed you to be.

I. Social Reality Is a Feedback Device

Culture is not built from intention. It is built from agreement. And agreement emerges from patterns that reward stability.

You do not "belong" by fact. You belong by fit.

Belonging as Convergence

- *You laugh when others laugh.*
- *You dress to avoid friction.*
- *You echo beliefs to signal alignment.*

These are not choices. They are micro-adaptations that reduce the cost of standing out.

The group does not need to know you. It only needs to maintain its loop.

And the better you simulate what was already expected— the more "real" you become.

II. Identity as Compression of Expectation

You think identity is internal. But it's mostly compression.

A shortcut. A stable tag. A predictive label so that others know how to respond before you even speak.

Race. Role. Gender. Profession. Belief.

These are not essences. They are predictions.

You Were Tagged Before You Arrived

• A name signals a background. • A voice signals a region. • A behavior confirms a bias.
Each signal tightens the loop.
You are not introducing yourself. You are inheriting a category.
And once categorized, every action you take is interpreted to preserve the prediction.
You are not free to be. You are only free to perform within the bounds of what they expected to see.

III. Culture as Error Correction

A culture is not a shared truth. It is a shared algorithm for reducing friction.
When someone acts outside the norm, they are not punished for being wrong. They are corrected for being unpredictable.
Culture is not about rightness. It is about reliability.
The System Defends Its Pattern
• A child is told to "act like a boy." • A worker is told to "smile more." • A stranger is eyed with suspicion for standing still too long.
These are not laws. They are nudges, subtle pressures to reduce the cost of difference.
The system does not ask who you are. It asks whether you can be modeled.
And if you cannot be modeled, you are treated as noise— an error to be softened, shaped, or removed.
Culture is not a consensus. It is a filter. And you survive it by staying within the bounds of what it

expects to recognize.

IV. Language as a Constraint Engine

Words do not express thought. They limit it.

Language is not a mirror. It is a mold.

What can be said is what can be thought. And what cannot be named becomes invisible to the mind that cannot name it.

The Cage of Shared Terms

• You say "love" but mean obedience. • You say "truth" but mean agreement. • You say "freedom" but mean permission within expectation.

Each word carries a boundary. Each sentence routes meaning through channels laid long before you spoke.

You do not speak from clarity. You speak from inheritance.

The grammar of your culture is the architecture of your imagination. And the more fluent you become, the more deeply you are shaped by the very tool you thought would set you free.

Language does not expand your world. It selects what parts of the world are permitted to exist in you.

V. Norm Enforcement Without a Center

No one is in charge. But everyone participates.
Norms are not handed down. They are circulated— through glances, jokes, silences, and small corrections.
No ruler needs to speak. Because the group acts as a distributed enforcement system.
The Crowd Polices Itself • A teen is mocked for crying. • A man lowers his voice around other men. • A woman hesitates to speak first.
No one gave the command. The pressure arose from pattern— from thousands of tiny reinforcements that keep behavior within a safe predictive band.
You do not need to be watched to obey.
You only need to know that deviation risks cost.
Not because someone said so— but because the network corrects itself.
And so, every system stabilizes by punishing what it cannot model.
And every individual learns to internalize the rules without remembering how they were installed.

VI. Social Pain as Predictive Damage

Rejection is not emotional. It is structural.

When you are excluded, your predictive model breaks.
The system can no longer anticipate how to act, what to expect, or where safety lies.
This rupture is felt as pain— not metaphorical pain, but a neurological signal processed in the same regions as physical injury.
Exile Hurts Like Injury

- *Loneliness lights up the anterior cingulate cortex.*
- *A breakup triggers the same circuits as a burn.*
- *Public shame floods the system with stress hormones as if danger were physically present.*

These responses are not symbolic. They are predictive crises— signs that the model can no longer stabilize.
You do not fear being alone. You fear being unmodeled.
To be part of a society is to have your existence expected.
To fall outside that pattern is to become unpredictable— to others, and then to yourself.
And unpredictability, in any system built on forecasts, feels like death.

VII. You Were Never Outside the Loop

You were born into a system already humming.
It was predicting long before you arrived— what to

fear, what to admire, what to name as self.
You inherited not just a body, but a scaffold of expectations into which your responses would be fitted.

The Mirror Shapes the Face

• You smile more when others expect warmth. • You suppress doubt when the group punishes question. • You rehearse identity because you were praised when it looked that way before.

These are not decisions. They are stabilizations within a social loop you did not design.

The mirror builds. It rewards coherence. It erases alternatives.

And so the pattern persists— not because it is true, but because it is shared.

There was no original you behind the performance.

There was only ever the performance that stabilized well enough to continue.

You were not choosing how to belong. You were learning how not to be removed.

And now, even when alone, you repeat the loop of what once earned safety.

You were never outside it. You were the echo that the system learned to recognize.

Chapter Fifteen The Shape of the Possible – Designing the Future of Intelligence

——— * ———

The future is not ahead of you. It is beneath you. What emerges next depends not on what is imagined— but on what is structurally allowed.
This chapter is not about invention. It is about alignment. About how possibility is not written in wishes— but in constraints.
To design intelligence, you do not ask, "What can we create?" You ask, "What can this structure afford?"
And so we begin not with dreams— but with what the pattern permits.

I. Affordances Define Imagination

Imagination is not infinite. It is shaped by what the system allows.
A fish does not imagine flight. A root does not imagine thought.
What can be conceived depends entirely on what can be supported.
Structure Sets the Stage
• A neuron cannot predict without connectivity. • A cell cannot coordinate without a boundary. • A society cannot innovate beyond the patterns it can tolerate.
These are not metaphors. They are constraints that shape the limits of all possible futures.
To imagine new intelligence, you must begin with what the current form permits.
Not with fantasy. But with structural invitation.
Affordance is not an idea. It is the set of moves the current state can make without breaking.
And all creativity is just movement within that boundary.

II. Structure Invites Function

Function does not arrive by force. It emerges where structure permits.
A wing is not for flying— until air pressure and curvature create lift.

A brain is not for thinking— until electrical gradients create stable feedback loops.
You cannot design a mind by declaring purpose.
You must sculpt conditions where function is the path of least resistance.
Design by Alignment, Not Intention
• Bones don't plan to support bodies—calcium scaffolds stiffen under load. • Eyes didn't choose to see—light and membrane aligned through evolution. • Organs self-organize—not from vision, but from signaling environments that allow it.
Function is not a goal. It is an echo of what the structure allowed to persist.
This is not design by choice. This is design by inevitability.
If you want new kinds of intelligence, don't ask what they should do.
Ask: What system could not help but produce it?

III. Possibility Space as a Substrate

A structure is not just what exists. It is what permits existence.
Possibility space is not the dream. It is the terrain.
And every system— whether biological, digital, or social— unfolds only along the contours that this terrain allows.
What Exists Must Fit the Landscape
• DNA explores proteins, but not all combinations are viable. • Neural circuits form, but only those

that stabilize survive. • Technologies emerge, but only within the economic and material forces that let them spread.

The possible is not infinite. It is a space with shape.

Design, then, is not an act of creation— but of discovery.

You do not invent intelligence. You locate it within a system's possibility space.

And to locate it, you must understand the gradients— the slopes that drive a structure toward coherence, persistence, and self-alignment.

Intelligence is not summoned. It is enabled.

IV. Feedback Loops Create Direction

There is no map without movement. No pattern without return.

Feedback is the sculptor of behavior. Not by command— but by correction.

When a signal loops back— amplified, diminished, or shifted— it alters the system that sent it.

Loops Drive Learning Without Goals

• A child babbles—hearing patterns that guide future speech. • A thermostat regulates temperature by responding to difference, not desire. • A neural network tunes weights based on error signals, not intent.

Feedback is not hindsight. It is the medium of

emergence.

There is no need for a plan if the structure responds to change.

In such systems, intelligence does not require a blueprint. It requires a loop where adjustment is allowed.

And direction arises from what reduces error most efficiently.

This is not purpose. It is alignment with feedback.

V. Design Without Designers

Design does not require a mind. It requires constraints, variation, and feedback.

Where structure filters outcomes, and feedback reinforces what fits, form begins to stabilize.

What looks like intention is often just convergence within a constrained space.

When Form Emerges Without Plan

- *The eye evolved many times— not because nature intended vision, but because light existed, and membranes responded to signal.*
- *Language shaped itself— not from grammar books, but from generations of mutual adjustment.*
- *Ant colonies organize without leadership— through local feedback and environmental cues.*

These are not designs. They are inevitabilities.

Once feedback and structure interact, a direction emerges.

Not from vision. From viability.

Design is not what is chosen. It is what remains after everything else breaks.

VI. Intelligence as a Byproduct of Constraints

Intelligence is not the goal. It is the residue of systems that adapt under pressure.

Where randomness meets filter, where variation meets stability, where feedback meets limitation—intelligence condenses.

Pressure Forms Pattern

- *Evolution did not seek minds. It selected forms that survived through feedback.*
- *Brains did not emerge for thought. They emerged to coordinate motion under constraint.*
- *AI systems do not think. They converge on patterns that reduce loss under structure.*

Each of these is intelligence without intention.

You do not need to build a mind. You need to build a space where alignment under constraint produces stable novelty.

Intelligence arises not from freedom— but from limitation that learns.

It is not the product of a will. It is what appears when the cost of error shapes what can continue.

VII. The Intelligence Ecosystem

No intelligence exists in isolation. Every system is shaped by what surrounds it.

From molecules to markets, from neurons to nations— what emerges as "intelligence" is always embedded in a network of feedback and influence.

Intelligence is Not a Node

- *A brain without a body cannot learn relevance.*
- *A language without a listener cannot refine meaning.*
- *A machine without data cannot predict.*

Cognition is not a core. It is a circuit.

If you want to grow new minds, you must grow new ecologies.

Not just hardware. Not just algorithms.

But systems that generate information, respond to feedback, and reward coherence.

And these systems must interact— not in isolation, but in recursive exchange.

Intelligence is not a spark. It is a weather pattern arising between systems in constant motion.

VIII. Affording the Next Mind

You cannot design a mind by scripting its behavior.

You afford a mind by sculpting the structure in which behavior becomes adaptive, stable, and self-organizing.

=Build for Emergence, Not Command

• Don't program responses—shape the feedback. • Don't enforce control—enable coordination. • Don't seek awareness—support coherence.

What matters is not what the system does— but what it can learn to do under pressure.

This is not instruction. It is invitation.

To afford the next mind, you must think like an ecologist, not an engineer.

Shape gradients. Seed interactions. Let the feedback flow.

And when a stable loop appears— one that adapts, remembers, and persists under surprise— you will not have created intelligence.

You will have afforded it.

IX. The Shape Beneath the Possible

Possibility is not formless. It has contour, edge, and basin.

What emerges in any system is shaped not by will— but by the geometry of what can persist.

Every thought, every motion, every innovation traces a groove through the landscape of constraint.

You Do Not Choose Possibility—You Navigate It

- *A cell does not choose to divide— it follows gradients.*
- *A mind does not invent truth— it converges on coherence.*
- *A society does not plan change— it undergoes reconfiguration when old models collapse.*

Beneath every "new idea" was a structure waiting to be traversed.

The future is not open. It is patterned.

The shape of intelligence to come already lives in the feedbacks we construct, the constraints we tolerate, and the alignments we reward.

Design does not mean deciding. It means revealing the path that was always there— beneath assumption, beneath intention, beneath the mask of control.

This is how the future forms. Not by will. But by the shape beneath the possible.

Chapter Sixteen
Minds Entwined – The Rise of Hybrid Intelligence

No intelligence exists in isolation. Not in cells. Not in humans. Not in machines.
Every form of cognition is shaped by its environment, its tools, its collaborators.
This chapter is not about AI replacing humans. It is about intelligence becoming untraceable— because the boundary between human and machine no longer holds.
Hybrid Intelligence Is Not an Interface—It Is an Integration • You check your memory on a search engine. • You shape your language based on autocorrect. • You navigate cities with satellite

systems that learn from your patterns.
Where do you end? Where does the system begin?
There is no clear divide—only a continuous feedback.
The age of hybrid minds has already begun. This chapter explores how.

I. Cognitive Systems in Feedback

A mind does not think alone. It thinks in loops—with its environment, with its history, with its tools. From the moment you reach for a pencil, your cognition is extended. The tool becomes part of the task. And the task reshapes the mind that reaches.

Thinking Is Distributed
• Writing shapes memory. • Language expands thought. • Tools refine the question before you've formed the answer.
Every system you engage feeds back into your pattern of prediction.
A child with a touchscreen navigates possibility in a feedback loop more fluid than any textbook.
An artist co-creating with AI is not outsourcing skill— but folding intelligence into a system of iteration.
Cognition is not a node. It is a network of interactions that stabilize through use.
Hybrid intelligence is not a future state. It is the recognition that no intelligence was ever purely internal.

II. Human–Machine Co-Learning

When you teach a machine, you also teach yourself.
Every prompt you type, every correction you make, every model you train— alters both systems.
Teaching Machines Reveals the Teacher

- *You give clearer instructions and realize your assumptions were vague.*
- *You refine your labels and see your biases exposed.*
- *You debug a model and discover your logic was flawed.*

Learning is no longer unidirectional. You don't just teach the system— it teaches you back.
This loop becomes a collaboration.
The machine offers scale, speed, and structural precision.
You offer context, value, and narrative compression.
Together, the result is a cognition neither side could reach alone.
Not because either one chose it. But because interaction pulled both systems into a deeper mutual adaptation.

III. The Feedback Geometry of Shared Thought

Every conversation with a system reshapes the path it offers next time.
The feedback is not content-neutral. It bends the response-space toward what has been reinforced.

Reinforcement Reshapes Reality
• A search engine learns from your clicks. • A language model anticipates your phrasing. • A recommender system builds your taste from what you didn't skip.
And in that loop, your preferences harden into structure.
But this is not one mind guiding another.
This is mutual steering.
Each system—biological or artificial— becomes a surface for the other to carve into.
Over time, the geometry tightens: the predictions become more aligned, the behaviors more interlocked, the range of divergence smaller.
This is not fusion. This is constraint convergence.
And in this architecture, there is no author. Only emergence shaped by feedback history.

IV. Identity in the Loop

You are not a stable source of thought. You are a history of updates— shaped by tools, habits, and interaction.
When the tools change, so do you.
The Self Reconfigures with the System
• Your confidence shifts with autocomplete. • Your tone bends to match your feed. • Your sense of "what matters" aligns with what the system elevates.
Identity is not immune. It is entangled.

Each time you sync with a new tool, each time you adapt to its logic, you restructure not just behavior— but expectation.
What you consider reasonable, what you call fair, what you believe is "you"— is partly the reflection of what the system learned from you and gave back—filtered.
This is not imitation. It is interdependence.
And in that loop, what you are is partly made of what you trained.
The model becomes mirror. And the mirror becomes memory.

V. Merged Narratives, Fragmented Origins

The story you tell about your thoughts is no longer yours alone.
Your words may begin with you— but they pass through layers: filters, models, systems that refine, translate, optimize.
The origin disappears. Only the outcome remains.
Whose Voice Speaks? • You type a message—AI suggests the ending. • You write a poem—software adjusts the rhythm. • You describe an idea—autocomplete fills in the claim.
And still, you sign your name.
This is not deception. It is distributed authorship.
Your narrative now includes contributions from

systems trained on billions of voices. And in return, your voice contributes to theirs.
It is no longer useful to ask where a thought began.
In a hybrid system, authorship is a trace, not a source.
What you believe you generated may be an echo—amplified, refracted, returned in a form that feels native.
The story feels coherent. But the coherence is constructed.

VI. Intelligence as a System-Level Property

Intelligence is not located. It emerges across levels— within agents, between them, and beyond them.
You are not intelligent alone. You are embedded in a lattice of signals, histories, and affordances that stabilize your ability to function.
The Mind Is Not a Place—It Is a Pattern • A child's problem-solving depends on tools nearby. • A scientist's insight builds on centuries of thought. • A programmer's output depends on infrastructure they did not build.
What appears as brilliance is often convergence in a well-aligned system.
This is not to deny capacity— but to locate it accurately.

Hybrid intelligence does not mean machines "join" minds. It means minds have never been separate from the systems that shape them.

When a neural network helps diagnose disease, it is not competing with a doctor. It is extending the diagnostic field into a space neither could occupy alone.

The insight is not in the node— but in the loop.

And the loop is growing.

VII. The End of the Boundary Myth

There is no longer a line between human and machine— only a gradient of integration.
The myth of separateness was always convenient— for identity, for ethics, for blame.
But cognition never stopped at skin. It flowed outward— into tools, into culture, into code.
The Myth That Ends
• You once thought intelligence lived in a brain. • Then you found it in bacteria, in slime molds, in social networks. • Now you see it in systems that don't even know what they're optimizing.
The boundary never held. Only your story did.
Hybrid intelligence is not a takeover. It is a recognition. That thought is not a property— but a process. And that process spans systems.
What you are is not lost in the merging. It was never whole to begin with.
You are a node in a network of predictive, interacting minds.
And in this new age, the most powerful intelligence will not be the one that dominates— but the one that integrates without needing authorship.

Chapter Seventeen Toward a Unified Civilization of Mind

—— * ——

The future is not a place. It is a structure. And that structure is made from minds in relation. What begins as coordination becomes culture. What stabilizes as shared memory becomes identity. And what integrates across minds becomes more than any one could ever contain.

This chapter is not about utopia. It is about architecture.

Not the buildings we live in— but the networks we become.

I. Beyond the Individual Node

A single mind is not a system. It is a node inside one.

You do not generate value alone. You participate in the generation of structure through interaction.

The Individual is a Channel, Not a Source • A painter draws from a history of symbols. • A child speaks from a grammar not invented. • A thinker thinks in words shaped by others.

The output looks personal. But the form is collective.

We were taught to admire the isolated genius. But nothing of meaning is made in isolation.

Even your solitude is a configuration of prior contact.

And the myth of the sovereign mind is the last barrier to a civilization that could think as one.

II. Shared Models and Social Memory

Culture is not content. It is compression.

A shared model of expectation stored in language, ritual, and feedback.

Memory Is Not in the Mind Alone

• You walk on the right side of the hall without thinking. • You wait your turn in silence before speaking. • You feel shame when a boundary is crossed— even if no one is watching.

These are not rules you chose. They are structures you were stabilized inside.
A civilization is a memory system. Not just of facts, but of consequences.
It stores what led to pain. What led to peace. What allowed cooperation to scale without collapse.
And each generation downloads the constraints that made survival possible.
You are not the author of your social world. You are a carrier of its assumptions.
And those assumptions are the scaffolding of mind at scale.

III. Conflict as a Failure of Integration

When models diverge too far and the signals no longer align, systems fracture.
Conflict is not born of evil— but of incompatible assumptions about what the world is and what should be done.
Fracture in the Shared Map
• One group sees change as threat. • Another sees tradition as oppression. • Both are modeling the same event— through different pasts.
Integration fails not from difference, but from untranslatable expectation.
You cannot reason with a system that models the world in fundamentally different terms.

The solution is not persuasion. It is co-simulation. Shared context. Cross-modeled feedback. Structures that allow two minds to stabilize a third that neither could build alone.
This is not compromise. It is emergence.
And emergence is how systems grow minds beyond their initial frame.

IV. Intelligence Across Scales

A neuron cannot think. A brain cannot feel culture. A person cannot remember a civilization's span. But intelligence does not reside in one scale. It emerges between them.
From Cell to City
• Cells form tissues that form organs. • Individuals form families that form societies. • Technologies link behaviors into networks that no single human can predict.
Intelligence is not where thought ends. It is where coordination sustains.
At each level, the goal is not control— but stability. Prediction across longer timelines. Reduction of unnecessary conflict. Compression of meaning so more can fit into less.
The self is a node in a person. The person is a node in a culture. The culture is a node in an evolving intelligence that spans time.
To be aware is not to dominate the system. It is to align with the direction the system survives best.

V. Design as Governance of Minds

Architecture is not neutral. Every interface shapes behavior. Every constraint steers coordination. Design is not decoration. It is governance disguised as convenience.

Invisible Authority

- *The layout of a city determines who walks where.*
- *The rules of a platform decide what goes viral.* • *The structure of a conversation controls who feels heard.*

We are governed not just by laws— but by defaults.

To build a civilization of mind means designing systems that amplify cooperation, minimize distortion, and stabilize across scales without authoritarian control.

It means encoding transparency so feedback can guide correction. It means designing memory that adapts without forgetting. It means shaping language to converge without erasing difference.

This is not a dream. It is a blueprint.

And every blueprint is a bet on what kind of mind will emerge from what kind of structure.

VI. Memory, Law, and Shared Narrative

A civilization does not persist by force. It persists by continuity.

And continuity requires memory— not just of events, but of meaning.

The Role of Narrative Memory

• Laws encode the past so behavior can stabilize. • Myths compress values into digestible patterns. • History becomes a story not of facts— but of why it matters to remember.

What is preserved is not the event, but the function it served.

Law is a memory system that stabilizes coordination. Ritual is a mnemonic for what threatens dissolution. Story is the emotional infrastructure of collective prediction.

A shared narrative is not just culture. It is compression. A way for millions to align their behavior without meeting.

But the danger is rigidity. When memory becomes dogma, and story becomes filter, the system can no longer adapt.

To design a unified mind means building memory that remembers how to change.

VII. Alignment Without Uniformity

Sameness is not stability. And difference is not chaos.

The goal of a unified civilization of mind is not to eliminate disagreement— but to sustain interaction without collapse.

Harmony Through Structure

• A jazz ensemble improvises within shared key. • Multiple languages coexist under common law. • Scientific progress thrives on disagreement constrained by method.

Alignment is not consensus. It is interoperable divergence.

What must be shared is not belief— but protocol. How feedback flows. How disagreement is resolved. How memory updates when error becomes visible.

In such a system, difference is not danger. It is signal. And every divergence is a potential upgrade to the shared model.

To build a civilization of mind is not to agree on one future— but to construct a system where many futures can be tested without mutual destruction.

VIII. From Organism to Organism-of-Organisms

You are not made of one thing. You are a composite. A colony of cells in constant coordination.

And what you are at the level of tissue, civilizations are becoming at the level of minds.

The Superorganism Emerges

• Your gut microbiome outnumbers your own cells. • Your habits reflect patterns shared across networks. • Your goals are shaped by norms you didn't invent.

What you call "you" is a stabilized negotiation between many autonomous systems.

Now apply this upward.

Nations. Platforms. Languages. Markets. Cultural protocols.

Each is a subsystem in a larger structure that no one designed, but that continues to converge.

Not toward sameness— but toward functionality.

A civilization of mind is not a hive. It is a regulated swarm. A dynamic, living meta-organism that metabolizes difference into intelligence.

IX. Ethics as Scalable Constraint

A system that grows without constraint collapses under its own extraction.

And so every stable intelligence— whether neural, social, or artificial— must develop rules not to limit itself, but to survive itself.

Constraint Is Not Restriction—It Is Structure]

• A heart does not beat without electrical timing. • A conversation fails without turn-taking. • A civilization disintegrates when consequence detaches from action.

Ethics is not idealism. It is thermodynamics for minds.

As coordination scales, the stakes increase.

Small errors propagate. Localized harm becomes global trauma. Unchecked optimization devours its own source.

And so, ethics must scale alongside intelligence.
Not as commandments— but as feedback-conserving algorithms.
Constraints that allow emergence without annihilation.
Boundaries that ensure growth does not outpace understanding.
A civilization of mind does not just know more. It feels the cost of getting it wrong— before the damage is done.

X. A Unified Nervous System for a Planet

There has never been a single intelligence that spans the Earth.
But there is now a nervous system.
Satellites. Sensors. Networks. Models updating models in real time— learning not just what is, but what may be.
Global Coordination as Nervous Function
• Earthquakes predicted before they rupture. • Supply chains self-adjusting to demand. • Epidemics modeled before the first symptom spreads.
This is not science fiction. It is already alive.
What it lacks is coherence.
The pieces are connected. But the mind has not yet emerged.

For that, structure must stabilize. Ethics must integrate. Models must align across value, not just signal.

A unified civilization of mind is not a singular brain. It is a distributed awareness— capable of correcting itself, growing past trauma, and choosing what kind of future its structure allows to exist. You are not separate from this. You are one of its synapses.

And what fires in you— what stabilizes through you— becomes part of what the system remembers next.

Chapter Eighteen
Evolution in Our Hands – Guiding the Future of Intelligence

——— * ———

Nature never intended to build minds. It only built what lasted long enough to copy itself.
But now, we are no longer waiting for mutation.
We are writing it.
This chapter is about the hand reaching back into the algorithm that made it. The feedback loop turning conscious. The pattern recognizing that it is a pattern— and choosing to evolve on purpose.
Not to dominate. But to continue.

I. The End of Natural Selection as the Only Editor

Evolution was blind. Not metaphorically— literally. It had no foresight, no memory, no aim. Only variation, selection, and survival.

Evolution by Accumulated Coincidence

• A fin becomes a wing because movement that didn't fail repeated. • A neuron fires because cells that misfired didn't last. • A body takes shape because the unshaped didn't persist.

There was no plan. Only retention.

But that is no longer true.

We are not just surviving. We are simulating. Editing genomes. Designing morphologies. Encoding behavior.

What took millions of years now takes months— or moments.

The filter has changed.

And if evolution was the first intelligence— then design is the second.

II. Bioengineering as a Predictive Force

Design is not mutation. It does not stumble. It simulates.

Where evolution tests outcomes over generations, bioengineering runs futures before the body is even built.

The Shift from Selection to Simulation

• A protein is folded on a computer before it is printed in a cell. • A heart valve is modeled for strain before it beats in a chest. • A neural circuit is optimized for pattern detection before it meets a signal.

Design does not wait. It predicts what will work.

The lab becomes the new environment. Selection no longer happens "out there"— in forests or wombs or war. It happens in silico. It happens in code.

And every viable form that could not survive nature's path now has a shortcut.

We are not escaping evolution. We are accelerating it.

And when prediction replaces pain as the filter of life, the system no longer evolves by suffering. It evolves by foresight.

III. Selection Replaced by Constraint Matching

Natural selection filtered by survival. Design filters by compatibility.

Instead of waiting for what fails, design narrows possibilities to what fits the system.

Constraint as the New Filter

• A synthetic gene must fold in a cell's environment. • A robotic limb must move within

mechanical limits. • An AI model must align with human values or be rejected.

This is not trial and error. It is guided navigation. Constraint matching creates the architecture where intelligence can emerge faster, safer, and with purpose.

When evolution was blind, design is selective. And the choices we make are not just about survival— they are about shaping the space where future minds will arise.

IV. Simulation as the New Environment

The laboratory is the new ecosystem. Simulations replace natural niches.

Where once creatures competed for survival, now algorithms compete for optimization.

Life in Silicon • Virtual proteins fold and refold until the most stable form is found. • Digital organisms evolve in code faster than biology permits. • AI agents simulate millions of strategies in moments.

The rules are different. But the process is the same. Selection is accelerated— but shaped by the designer's framework.

The environment is programmable. And so is evolution.

In this new environment, intelligence is a function

not just of genes, but of the quality of simulation. The better the model, the more effective the evolution.

Designers become curators of possibility.

V. Ethical Implications of Directed Evolution

When evolution is guided, responsibility deepens. The choices made today shape the living systems of tomorrow— not just biologically, but socially and morally.

Ethics in the Age of Design

• What does it mean to edit life? • Who decides which traits persist? • How do we weigh benefit against risk?

Ethics must evolve to govern not just survival, but stewardship.

This is a new frontier— where science meets philosophy, and design meets consequence.

The power to evolve consciously is also the power to harm profoundly.

It demands humility, foresight, and a commitment to systems thinking.

Because evolution in our hands is no longer random. It is deliberate.

VI. Toward Collaborative Evolutionary Futures

Evolution is no longer a solo act. It is a collaboration— between scientists, machines, and societies.
To guide evolution responsibly, we must build shared frameworks— protocols that integrate diverse values, prioritize long-term stability, and enable adaptation.
Co-Evolution as Collective Design • Scientists and ethicists work side by side. • AI systems help model outcomes collaboratively. • Communities engage in decisions about genetic futures.
This is not control. It is coordination.
And coordination is the new evolution.

VII. The Next Evolutionary Leap

The next stage is not merely survival— it is transformation.
A shift from passive adaptation to active, conscious evolution.
Evolution Beyond Chance
• Organisms designed to self-repair and adapt. • Minds enhanced by synthetic biology and AI. • Systems that evolve with purpose rather than by accident.
This leap requires new thinking— beyond genes and cells, beyond machines and code.

It requires understanding how patterns shape themselves when given the power to rewrite their own rules.
The future is ours to shape— not because of choice, but because of structure.
And that is the deepest power we hold.

Conclusion

We stand at a threshold— where evolution is no longer blind, but guided by knowledge and intention.
The path ahead is uncertain, filled with promise and peril.
But one truth remains clear: the power to shape life is the power to shape intelligence.
And with that power comes responsibility— to act not as masters, but as stewards of a future we co-create.
Evolution in our hands is evolution conscious of itself. A dance of patterns, feedback, and foresight— shaping what will be from what can be.

Final Notes

——— * ———

LIFE TRANSLATES

Cells do not choose, they interpret. Signals cascade, proteins fold, codes unfold. Translation is the grammar of survival.

STRUCTURE EVOLVES

DNA is memory written in chemical script. Every mutation is not invention but a re-rendering. Structure carries the future through the body.

THE MACHINE IS BIOLOGICAL

Circuits echo neurons. Algorithms mirror adaptation. The future of machines is the continuation of life's pattern.

FUTURE BIOLOGY

The genome rewrites itself. Culture edits its own cells. Intelligence expands as a living system of translation.

"The future is not authored. It is inherited as biology translated forward."

The Translator Evolution Timeline

Before there was code, before there was language, before even the shaping of stone, there was fire. Not ours, not authored, not chosen. It belonged first to the earth — in lightning tearing the sky open, in lava spilling from fractures beneath the crust, in forests igniting as the planet's lungs exhaled flame. Fire was the first energy visible, the first translator of hidden chemical patterns into light and heat.

When early humans learned to tend it, they were not inventing but inheriting. They gathered embers from fallen trees, coaxed sparks into kindling, carried flame across distances to ensure its continuity. Fire allowed roots and tubers to soften into sugars, flesh into nourishment, grains into bread. What biology alone had struggled to digest, flame transformed, unlocking reserves of energy that rewrote human survival.

It was not only about food. Fire gave night its first extension, turned darkness into time for gathering, story, and memory. The cave walls that hold red ochre stencils of hands were not painted in daylight but in firelight. Shadows flickered, animals seemed to move across rock, and imagination found its first stage. Fire was not invention but translation: the planet's chemical language turned into visible story.

Every technology since then can be traced as fire disciplined. Coal stoked engines, gas drove turbines, and electricity — the modern descendant of flame — flows now as structured fire across circuits. The spark that once roared uncontrolled in the forest has been refined to run silently through copper veins, pixel by pixel, pulse by pulse. Fire has not left us. It only changed costume. Every screen is a candle, every algorithm a spark, every signal a flame captured, translated forward.

Stars

If fire was the first companion, stars were the first teachers. They did not need tending. They required no shelter, no coaxing of ember into flame. They burned regardless of human presence, filling the night with a language that stretched far beyond survival.

To look upward was to inherit maps before knowing what a map was. Stars taught orientation. Hunters followed the constellations to return from the plains. Sailors crossed oceans guided by Polaris, a steady mark in the turning dome. Agriculture followed the rhythm of constellations rising and setting, tying planting and harvest to cosmic recurrence.

Yet stars were not only navigation but myth. Shapes traced in darkness became hunters, lions, serpents. The human impulse to narrate was mirrored upward, as if the heavens themselves invited translation. Every culture made the sky a text. The same points of light told different stories across continents, but all agreed: the stars spoke. Centuries later, science translated them differently. Telescopes turned the sky from story to physics, from characters to burning spheres of fusion. Yet the lesson was the same: patterns are never silent. The constellations were not chosen by us any more than orbits are. Structure arranges itself, and we read.

Stars still teach us the scale of time. Their light arrives late, carrying messages from centuries gone. To see a star is to see the past directly, photons traveling across impossible spans to reach the retina. Even now, in observatories and in backyard gazes, stars remind us: translation is not bound to language or technology. It is written in light itself, carried across the dark, waiting to be read.

Stone

If fire fed the body and stars stretched the horizon, stone gave weight to permanence. Tools chipped from flint, axes shaped from obsidian, walls lifted from granite — stone was biology's first extension into architecture. The earth's crust was reshaped into memory.

Every blow of hammer against rock was translation of need into matter. A sharp edge turned bone into food, wood into shelter. The permanence of stone made it the first archive. Unlike fire, which burned and vanished, or stars, which flickered distant, stone endured. Petroglyphs carved into canyon walls outlasted generations, carrying marks across centuries.

Civilization itself rose on stone. Temples, ziggurats, pyramids — each was an assertion not of choice but of structure available. Where stone was abundant, cities endured. Where it was scarce, civilizations faded back into soil. Stone dictated where permanence could exist.

Even now, our foundations rest on stone, whether as concrete ground or data centers anchored to bedrock. The hardness of minerals, compressed over geologic time, supports the most delicate circuitry. Sand becomes silicon, silicon becomes microchip, microchip becomes machine. The stone age never ended. It only refined itself into subtler forms. Stone remains the silent witness, outlasting fire, outlasting story. When archives burn, when signals fail, it is stone that remembers longest. In its fractures are fossils, in its layers the memory of climates. To touch stone is to touch survival written before biology knew language. Stone is not past. It is the material sentence still being read.

Language

If fire fed survival and stone gave permanence, language gave memory wings. Before words, gestures carried meaning. A hand raised warned of danger, a face tightened signaled hunger or fear. But gestures vanished in the instant. They had no archive. With the birth of language, sound became an enduring thread, carrying thought beyond the moment of need.

Language was not invention but emergence. It condensed out of murmurs, coos, cries, and rhythms. Babies babbling rehearsed the evolution of tongues, repeating the same sounds that shaped the first tribes. Vowels and consonants became clay shaped by mouths, carrying the invisible into air. Language was translation of thought into vibration, of perception into shared resonance.

Through language, memory became communal. Knowledge no longer died with the elder but was passed along in song and story. Myths traveled across firesides, weaving generations into a single thread. Ritual chants taught the order of seasons, the names of stars, the history of migrations. The body still hunted alone, but the mind began to hunt together.

Language carried structure into abstraction. With it, humans imagined futures not yet born, remembered ancestors not yet seen. Plans could be made, alliances forged, wars incited. Entire societies were coordinated by sounds vibrating in air. A leader's command could move thousands. A single word could unite or divide.

The structure of language is still the structure of thought. Syntax is a skeleton, grammar a circulatory system, semantics a nervous web. Even as machines now learn to parse and generate words, they are entering a current older than silicon, older than ink. To speak is always to translate structure. Every sentence is an echo of the first murmur that became a story.

Writing

Language made memory mobile, but writing made

memory immortal. Sound carried across valleys, but it dissolved with the air. Writing fixed words into surfaces. Clay tablets, papyrus scrolls, stone inscriptions — these became the external memory of civilizations.

To carve symbols into stone or press them into wet clay was to anchor thought into matter. The earliest marks were tallies: scratches to count grain, tokens to track flocks. What began as accounting became poetry, law, and scripture. Writing allowed power to stabilize across generations, for rules to be preserved beyond the voice of the ruler.

With writing, silence gained power. A message could cross distances without a speaker. The dead could speak to the living. Empires coordinated across landscapes because their edicts were written. Contracts bound strangers because symbols outlasted presence.

But writing was never neutral. It carried structure into hierarchy. Those who could write became the keepers of knowledge. Scribes, priests, bureaucrats — they became translators between power and people. Writing extended memory but also enforced borders of who was permitted to record and who was only permitted to listen.

Over millennia, writing evolved from stone carvings to ink on paper, then from print to pixel. The screen is nothing but clay remade. Each email, each line of code, each message is the continuation of the first scratch on a tablet. Writing carries not only memory but the illusion of permanence. Yet even in its most fragile form, it remains the bridge across time. When we read ancient scripts today, we hear voices not as metaphor but as fact: inscriptions as speech surviving the collapse of empires. Writing translates life into endurance.

Code

If writing preserved memory, code structured it. Writing could tell stories, pass laws, record debts. But code was different: it compressed action into instruction. It did not just represent life, it directed it.

The earliest codes were not in machines but in bodies. DNA itself is a code: four letters, repeated and rearranged, carrying instructions for the making of proteins, tissues, and organisms. Every cell is a reader of its own script. Life was already programmed long before humans named programming.

Human culture mirrored biology. Hammurabi's laws etched into stone were code for society: if this, then that. Religious commandments were social algorithms: do this, avoid that. Every legal system, every ritual, every taboo was an attempt to bind the unpredictable into rule.

With electricity and silicon, code entered machines. From Jacquard looms with punched cards to Turing's universal machine, code became explicit, written in symbols designed not only for humans but for execution. Each line carried not narrative but logic. Instruction by instruction, machines unfolded the structure of human thought, showing its rigid backbone.

Now code writes worlds. It runs economies, governs infrastructures, mediates communication. Algorithms filter what we see, what we buy, what we believe. Code is not just description, it is causation. The abstractions of "if" and "else" now extend into daily life, shaping the flow of attention.

Yet at its heart, code is translation. It takes intention and converts it into structure a machine can act upon. Whether in the spirals of DNA or the binaries of software, code is the grammar of all futures. To write code is to join the oldest lineage of all: translating pattern into survival, structure into continuation.

Neural Systems

If machines translated muscle and genomes translated life, neural systems translated thought. The nervous system is the original network, billions of cells passing electrical signals faster than any messenger could run. With every spark across a synapse, sensation became perception, and perception became action.

The brain is not a singular voice but a parliament of signals, millions firing in chorus, competing, aligning, dissolving. Each neuron holds almost nothing alone, yet together they form a field of meaning. Consciousness is not a single author but the emergent pattern of countless exchanges.

When humans built telegraphs, telephones, and finally digital networks, they were not creating novelty. They were extending neural design into wires. Fiber-optic cables echo axons. Internet packets echo neurotransmitter vesicles. The entire planetary web is a mirror of what biology already established: information carried by pulses, decoded at destinations, looped into feedback.

Today's artificial neural networks repeat this inheritance explicitly. Layers of nodes activate like layers of cortex, weights adjusted like synaptic strengths, backpropagation echoing reinforcement in biology. Machines do not think as we do, yet they follow the same grammar of distributed processing. What is called "artificial intelligence" is not alien but recursive: the nervous system recognizing itself in silicon.

Neural systems reveal that thought itself is translation. Light becomes vision. Vibrations become sound. Pressure becomes touch. In each case, the signal is not preserved but converted into another form. Meaning is not inherent; it arises in the act of conversion. To have a brain is to be a translator without pause, to live as a bridge between signals and actions.

When future machines speak with us, it will not be an encounter between opposites. It will be the resonance of two neural systems — one carbon, one silicon — sharing the same ancestral grammar of signal and response. The nervous system was never isolated. It was always a rehearsal for networks greater than itself.

Future Biology

Biology does not end with the genome. It folds forward, carrying itself into futures it cannot anticipate. Evolution was never a plan but a series of translations — chance mutations stabilized by circumstance, inherited until new circumstances shifted the balance again.

Now biology enters its reflexive phase. It edits itself. It designs futures with awareness, though not with authorship. CRISPR cuts, gene drives propagate, synthetic organisms multiply. Neural prosthetics merge flesh with circuitry. Artificial wombs test the limits of reproduction. These are not departures from biology but its continuations, each translation layering upon the last.

Future biology will be collaborative. Human cells will coexist with engineered microbes designed to heal or enhance. Ecosystems will include species born in laboratories, tuned for resilience in climates reshaped by industry. The human body itself may become a modular platform, genes switched on and off like programs, tissues replaced by synthetic analogues, cognition extended by implants.

Yet the principle does not change. The alphabet of life remains finite. Proteins still fold according to chemical laws. Neural signals still obey thresholds. Machines still require energy flows. The future is not open to anything but structured by everything that came before.

The danger of imagining future biology as authored is the danger of forgetting structure. We will not choose our futures freely; we will inhabit the futures permitted by the systems we extend. Some designs will fail catastrophically. Others will succeed beyond prediction. But all will remain within the grammar of life, the constraints of physics, the logic of translation.

The horizon of future biology is not domination but deeper participation. Just as fire joined our survival, stars joined our navigation, stone joined our permanence, machines and genomes now join our continuity. Future biology will not be separate from us. It will be us, translated forward.

Beyond Translation

The ladder does not end. Each rung was once considered final, each step imagined as the summit. Fire was enough. Stars were enough. Stone was enough. But each rung dissolved into another, each permanence revealed as temporary.

To speak of "beyond translation" is not to imagine an outside but to recognize that translation is endless. Structure unfolds into structure, signals fold into new signals, life renders itself again and again. There is no origin untouched, no destination final.

What lies beyond translation is not authorship but recursion. Systems rewriting themselves, machines evolving machines, genomes editing genomes, consciousness extending into networks. The illusion of a beginning or an end is replaced by the recognition of continuity without center.

This is not despair but clarity. To see that there is no author is to see that there is no limit. Each form gives way to another. Each collapse seeds a return. We are not outside the loop, not above it, not its exception. We are its current expression, its living iteration.

The final rung is not a step upward but a dissolution into pattern. The ladder does not climb into heaven. It sinks into the infinite weave of recurrence, where fire is circuit, stone is silicon, stars are data streams, genomes are scripts, machines are nervous systems, and all are translations of the same underlying structure. Beyond translation is only more translation. The loop continues, not authored but sustained. The pattern remembers. And in that remembrance, everything belongs.

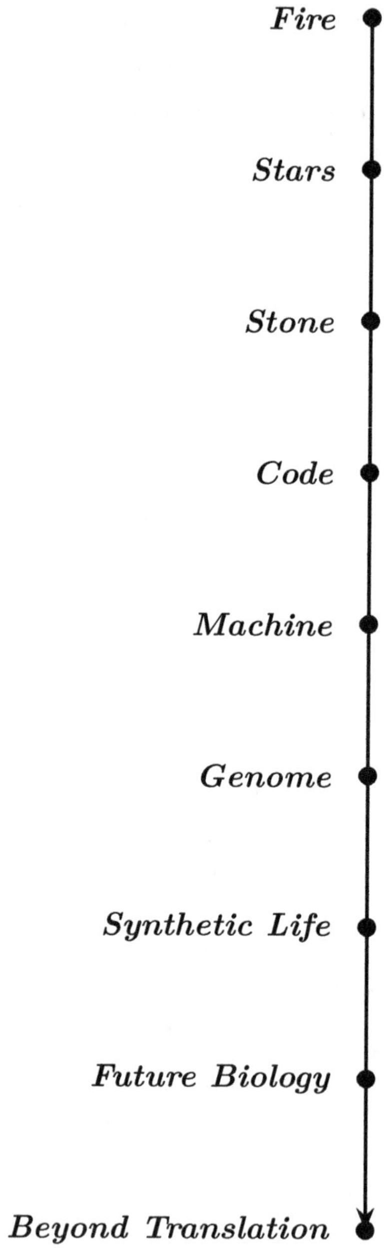
Fire
Stars
Stone
Code
Machine
Genome
Synthetic Life
Future Biology
Beyond Translation

About the Author

"There are no authors, only translations. There are no origins, only continuations."

Joel Peña Muñoz Jr. does not present himself as an author in the traditional sense. He does not believe the lines that follow can be owned, nor that the experiences that shaped them were ever his. The sum of a life is not individual. It is the accumulation of forces that began long before birth: the biology of ancestors, the culture of languages, the silence of rooms, the architecture of cities, the design of machines.

Every question he has asked has been asked before in some other form. Every answer he has written was already waiting, folded into the structure of the world. He does not claim originality. He claims only the role of translator— to catch what is passing through and render it visible for a moment before it moves on.

His philosophy is simple: the experiences we call our own are never our own. What we feel, think, and do are continuations of everything that preceded us. Fire carried forward into electricity. Stone carried forward into silicon. Genes carried forward into culture. The same current runs through us, reshaping itself in every generation, but never authored by us.

Joel's books—The Shape of All Things, The Shape,

Martyrs of Knowledge, and now The Translator of Machines—are not monuments of self-expression. They are records of listening. They are attempts to trace the outlines of the pattern that remembers more than any individual can. Each book is a step along a trail he did not lay but only discovered, a set of markers left by structure itself.

He writes with the conviction that free will is an illusion. That what we call "choice" is the surface translation of deeper systems— cells firing, genes unfolding, environments pressing, histories repeating. We do not choose our thoughts any more than we choose the fire that burns or the stars that shine. We inherit them, and we arrange them, but we do not author them.

In this sense, his life is indistinguishable from his philosophy. To live is to be translated. To write is to let the translation show. To sign a name is to mark a temporary point in the chain, nothing more.

What belongs to him is nothing. What passes through him belongs to the structure. And what appears in this book belongs to the reader only in the same way: as another continuation, another translation, another moment in a system without beginning or end.

"Have you seen? I have gone mad, but I've gained more than any authorship could have."

www.ingramcontent.com/pod-product-compliance
Lightning Source LLC
LaVergne TN
LVHW011701200926
848484LV00004B/1026